ANALOG VHDL

edited by

Andrzej T. Rosinski†
Institute of Electron Technology
POLAND

Alain Vachoux
Swiss Federal Institute of Technology
SWITZERLAND

A Special Issue of
**ANALOG INTEGRATED CIRCUITS
AND SIGNAL PROCESSING**
An International Journal
Volume 16, No. 2 (1998)

KLUWER ACADEMIC PUBLISHERS
Boston / Dordrecht / London

ANALOG INTEGRATED CIRCUITS AND SIGNAL PROCESSING

An International Journal

Volume 16, No 2, June 1998

Special Issue: Analog VHDL
Guest Editors' Andrzej T. Rosinski[†] and Alain Vachoux

Guest Editors' Editorial . . . . *A. Rosinski[†] and A. Vachoux*		1
Part 1: VHDL Use for Analog Modeling		
Discrete Approach to PWL Analog Modeling in VHDL Environment . . . . . . *J. Dąbrowski and A. Pułka*		3
Oversampling ΣΔ Analog-to-Digital Converters Modeling Based on VHDL . . . . *R. Baraniecki, P. Dąbrowski and K. Hejn*		13
VHDL Modeling of Optoelectronic Interconnect Networks . . . . *S. Koh*		23
Part 2: Experiences with an AHDL		
Hierarchical Analog Behavioral Modeling of Artificial Neural Networks . . . . *M. Ahmed, H. Haddara and H. Ragaie*		33
Fault Modeling and Simulation Using VHDL-AMS . . . . *A. J. Perkins, M. Zwolinski, C. D. Chalk and B. R. Wilkins*		53
Creative Methods of Leveraging VHDL-AMS-like Analog-HDL Environments. Case Study: Simulation of Circuit Reliability . . . . *S. GadelRab and J. Barby*		69
Part 3: Analog VHDL Issues and Solutions		
Analog Behavior Modeling and Processing Using Ana VHDL . . . . *L. Ye and H. Carter*		85
Analog and Mixed-Signal Extensions to VHDL . . . . *A. Vachoux*		97

Distributors for North, Central and South America:
Kluwer Academic Publishers
101 Philip Drive
Assinippi Park
Norwell, Massachusetts 02061 USA

Distributors for all other countries:
Kluwer Academic Publishers
Distribution Centre
Post Office Box 322
3300 AH Dordrecht, THE NETHERLANDS

Library of Congress Cataloging-in-Publication Data

Analog VHDL / edited by Andrzej T. Rosinski, Alain Vachoux.
      p.   cm.
   "A special issue of Analog integrated circuits and signal
processing, an international journal, Volume 16, No. 2 (1998)."
   Includes bibliographical references (p.).
   ISBN 0-7923-8160-2 (alk. paper)
   1. Linear integrated circuits--Computer simulation.   2. VHDL
(Computer hardware description language) I. Rosinski, Andrzej T.
II. Vachoux, Alain. III. Analog integrated circuits and signal
processing.
TK7874.A578   1998
621.39'2--dc21   98-16856
                                                                                                CIP

Copyright © 1998 by Kluwer Academic Publishers

All rights reserved. No part of this publication may be reproduced, stored in a retrieval system or transmitted in any form or by any means, mechanical, photo-copying, recording, or otherwise, without the prior written permission of the publisher, Kluwer Academic Publishers, 101 Philip Drive, Assinippi Park, Norwell, Massachusetts 02061

Printed on acid-free paper.

Printed in the United States of America

Editorial

This Special Issue of Analog Integrated Circuits and Signal Processing is dedicated to the application of the VHDL hardware description language to the design of analog and mixed-signal circuits and systems.

VHDL is an IEEE standard language that has been primarily designed for the description and the simulation of digital hardware systems. The first version of the standard has been released in 1987. A second version that corrects language inconstancies and includes new mechanisms has been released in 1993. Since its inception VHDL has gained considerable importance as it is today supported in all major Electronic Design Automation environments available in the market. It is also going beyond its original goals as it also efficiently supports automatic synthesis, formal proof, and testing. As such, VHDL is acting as a unified medium that supports a large part of the design process, from the abstract specifications to the logical implementation into gates.

Some attempts have been made to use VHDL for the description and the simulation of analog and mixed-signal circuits, with limited success however. The problem has its grounds in the discrete time event-driven semantics that are intimately part of the language. The continuous aspects that are specific to analog behavior have then to be re-expressed in a discrete way in order to obtain legal VHDL models. This approach converts the original analog component into a data-sampled model that may exhibit spurious effects that are only due to the conversion. For example, the manual discretization of time derivatives makes the VHDL model dependent on a numerical integration method and on a specific integration timestep. Also, the possibly very large number of events that could be generated during simulation may make this approach intractable in practice. Pushed to the extremes, this approach mainly uses VHDL as a programming language to write dedicated simulators, rather than to write models.

This apparently bad state of affairs should be anyway moderated by a couple of other facts. First, there are cases for which abstract, signal-flow like, models of analog components are accurate enough because they are used to validate a complete system. In the case of a mixed-signal system in which most of the functionalities are digital, too detailed models won't bring much more information in a system simulation, but will certainly increase the simulation time drastically. Second, data-sampled analog systems, such as switch capacitor circuits or oversampling data converters, already exhibit discrete time behaviors that may be naturally expressed in VHDL. Third, the near availability of the VHDL 1076.1 language (also informally known as VHDL-AMS), an extension of the VHDL language to support analog and mixed-signal systems, is promising to offer powerful means to efficiently write and simulate models that exhibit continuous behavior over time and amplitude.

The papers in this Special Issue are grouped into three logical parts that address key aspects related to the main topic of the Issue. The first part deals with the application of the VHDL language to analog and mixed-signal modeling. In the first paper, J. Dąbrowski and A. Pułka present a way to model analog systems at a functional level, i.e. neglecting bidirectional effects such as loading effects, with piecewise linear techniques that lead to the development of high-level macromodels that can be efficiently simulated in VHDL. The second paper, from R. Baraniecki et al., gives the details of the VHDL modeling and simulation of an oversampling sigma-delta A/D converter. The models are also high-level but they include imperfections such as hysteresis and slew rate. Reported simulation times increase anyway very rapidly with the number of decimator stages. In the third paper, S. Koh shows how it is possible to develop mixed-technology models in VHDL to support, Koh says, an optoelectronic interconnect simulation methodology. The approach is validated through the modeling and the simulation of an optoelectronic clock distribution network. All of these modeling approaches define a basic set of VHDL declarations that are included in specific VHDL packages. These declarations are required to represent analog waveforms in a discrete way and to apply appropriate operations on them. The second part of this Issue provides experience reports on the use of an analog hardware description language for specific applications. In the fourth paper, M. M. Ahmed et al. report on the use of the HDL-A language, a proprietary language from Mentor/Anacad that has some commonalities with the VHDL-AMS language, on the modeling and simulation of neural networks. The paper provides a detailed discussion on the development of a library of accurate electrical models of primitive cells such as a Gilbert multiplier. Next, A. J. Perkins et al., describe how to develop behavioral models of analog components that support fault simulation. They discuss the pros and cons of the application of a VHDL-AMS-like language (HDL-A again, actually) versus a more traditional SPICE based approach. The sixth paper, from S. M. Gadelrab and J. A. Barby, shows how a language such as VHDL-AMS may efficiently support very specific simulation capabilities that goes beyond what electrical simulators actually offer. The flexibility of the language allows to develop additional modules that use the existing simulation algorithms to realize, as a demonstrative example, circuit reliability simulation.

The third part of this Issue explores VHDL language extensions that aim to more efficiently support the modeling and simulation of analog and mixed-signal systems. The paper from L. Yeh and H. W. Carter presents the AnaVHDL language that is an academic effort to develop a minimal extension of the VHDL language to support the direct expression of differential algebraic equations (DAEs). This paper also provides insights into the underlying equation formulation methods that are required to simulate a model written in an analog hardware description language. Finally, the last paper from A. Vachoux presents the key aspects of the upcoming VHDL-AMS language that is the official IEEE proposal to extend VHDL to support analog and mixed-signal systems.

This Special Issue has a very long history behind it as the first call for papers has been issued in 1995 from the initiative of Professor Andrzej Rosinski from the Department of Microelectronics of the Institute of Electron Technology in Warsaw, Poland. Prof. Rosinski has been actively investigating the use of VHDL, particularly its application to the modeling of analog and mixed-signal circuits. Prof. Rosinski suddenly deceased in January 1997 and I would like to dedicate this Special Issue to his memory.

I would like to thank the reviewers for their effort and time to provide many useful comments that improved the quality of the papers. Special thanks to the Editor-In-Chief Mohammed Ismail for his long-standing patience and to Ms. Karen S. Cullen and all the staff at Kluwer Academic Publishers for their efficient handling of the production of this issue.

Alain Vachoux
Andrzej Rosinski†
Guest Editors

Discrete Approach to PWL Analog Modeling in VHDL Environment

JERZY DĄBROWSKI AND ANDRZEJ PUŁKA
Institute of Electronics, Silesian Technical University, Gliwice, Poland

Received August 8, 1996; Accepted June 5, 1997

Abstract. In this paper a discrete approach to analog modeling is presented. It is a functional-level, piecewise-linear (PWL) technique implemented in the VHDL environment. Since the models are based on some explicit formulas, fully behavioral architectural bodies have been proposed for them. Their most distinguishing features are discussed in detail. The models of practical circuits are illustrated with simulation results.

Key Words: PWL modeling, analog A/D modeling, mixed signal modeling, mixed A/D simulation, functional-level analog modeling, VHDL analog modeling

1. Introduction

While dealing with complex electronic designs, the problems arising with their analog parts cannot be easily avoided. Usually separate analysis of the extracted analog sub-circuits does not provide fully adequate timing results. High speed circuits, even purely digital, exhibit some analog properties having their origin in interconnection delays. On the other hand, also at higher levels of abstraction the need of more accurate analog modeling begins to emerge. In particular, the development of analog and mixed-signal application specific circuits (ASICs) has an evident impact on this process. The orientation aimed at a complex and unified treatment for analog or mixed A/D designs can be observed in available publications, e.g. [1,2]. The work towards VHDL-AMS seems to be the most meaningful in this field.

In this paper our attention is focused on modeling of analog units at the functional level. We use the piecewise-linear (PWL) approach [5,6], which makes it possible to consider analog models as discrete objects at the input and at the output as well. Consequently, no continuous simulation technique is required to handle our models. It means, that we do not follow the basic VHDL-AMS highlights in our approach. Instead, the analog models are represented in the discrete VHDL environment and the behavioral approach to create their architectural bodies is used. The main implementation issues are discussed in detail. We show the particular solutions for some practical analog units, like a low-pass filter or an integrator. The included simulation examples, obtained by the V-System software [9], validate the models.

2. Approach to PWL Modeling

Based on the basic algebraic operators and some inertial building blocks the synthesis of functional-level analog and/or analog-digital macromodels can be performed [6,11]. The inputs applied to these models are assumed to be PWL signals. Hence, the respective time responses can be obtained by explicit formulas. However, a special approximation procedure is used, which in fact replaces the original waveforms with the PWL segments, to enable further propagation of the smooth output signals in the PWL form. In this way, when a network is modeled, all the links between analog units take a form of the PWL signals.

For analog units two characteristics have to be modeled: the static (DC) and the transient one. While a single pole transient characteristic (response) is assumed, a model based on a first order differential equation takes the form of:

$$\tau \frac{dx}{dt} + x = f(x_{inp}) \quad (1)$$

where x_{inp} is the input signal, τ is the time constant and $f(\bullet)$ represents the static characteristic (e.g. of an amplifier or comparator) and it may be generalized into a multiple input case (e.g. for an adder).

If loading effects at the output must be accounted for, this model can be accomplished by adding a second stage:

$$\tau_0 \frac{dx}{dt} + y = x \qquad (2)$$

For a capacitive load C_0 we have $\tau_0 = (C_0 + C_s)R_s$, where C_s and R_s are respectively the internal output capacitance and resistance. Using a PWL signal $x_{inp}(t)$ and a PWL approximation for $f(\bullet)$, also a PWL signal $u(t) = f[x_{inp}(t)]$ is obtained.

Let $x(0) = x_0$ and $u(t) = u_0 + rt$, $t \in [0, t_{\max}]$. Hence, solving for (1) we have an explicit formula:

$$x(t) = (x_0 - u_0 + r\tau)e^{-t/\tau} \qquad (3)$$
$$+ r(t - \tau) + u_0$$

The main objective faced here is to get a PWL approximation of (3) to enable further propagation of the signal x in a linearized form.

For this purpose, first we split the time interval $[0, t_{\max}]$ into subintervals $[0, t_1]$, $[t_1, t_2]$, $[t_n, t_{\max}]$. For each subinterval $[t_i, t_{i+1}]$ a segment of PWL approximating signal x_{lin} is defined by its boundary points, that are assumed to lie on the original curve x. Hence we have: $x_{lin}(t_i) = x(t_i)$ and $x_{lin}(t_{i+1}) = x(t_{i+1})$.

In fact, given t_i, the t_{i+1} has to be found. To control the accuracy of this approximation, the Chebyshev measure may be used. It has been found the most advantageous to develop an efficient approximation algorithm described briefly below. Consequently, our objective can be formulated as an optimization task, that is to maximize the distance $d = t_{i+1} - t_i$ with some constraints and given t_i:

$$\underset{t_{i+1}}{\text{Maximize }} d : \{d = t_{i+1} - t_i, \ t_{i+1} \leq t_{\max}\} \qquad (4a)$$

$$p(t_i, t_{i+1}) = \underset{t \in (t_i, t_{i+1})}{\text{Max}} |x(t) - x_{lin}(t)| \qquad (4b)$$

$$p(t_i, t_{i+1}) \leq p_{\max} \qquad (4c)$$

where $p(t_i, t_{i+1})$ is the performance index and $p_{\max}$ is an arbitrarily chosen constant (maximum) allowed approximation error). During simulation it is important to calculate the subsequent points $(t_{i+1}, x(t_{i+1}))$

in an efficient way and possibly to avoid iterations, that all typical optimization procedures (e.g. Fibonacci search) are based on. Hence, we use an explicit algorithm to solve this problem [6].

In Fig. 1 a process of the PWL signal conversion is shown which is typical in the analog PWL modeling we use, and in practice only a few exceptions exist (e.g. modeling of the slew-rate effect in an amplifier) [6]. The first block in this diagram is usually the mentioned above inertial building block, whereas the second one is the approximator. The latter outputs the PWL segments, which are defined by their discrete boundary points.

The approximator works in two steps [6]. First, the time t_a is computed to find the range of an acceptable quadratic approximation for the original response. Based on the Taylor series expansion the third order Lagrange rest is taken to evaluate $t_a = \tau*(6\varepsilon_a/\text{abs}(x_0 - u_0 + r\tau)) \wedge 0.33$, where ε_a is an estimated value of the maximum allowed truncation error. In the second step an attempt to set the PWL segment is made, so that it possibly covers the full range of t_a (for the end point of the segment we would have $t_1 = t_a$). If the Chebyshev distance $p(0, t_a)$ between the PWL segment and the original waveform exceeds the prescribed approximation accuracy $p_{\max}$, then the boundary point t_1 has to be re-evaluated from the simple proportion: $t_1/t_a = \text{sqrt}[p_{\max}/p(0, t_a)]$. Otherwise the primary evaluation for t_1 holds. The next PWL segment can be found in the same way. Since the original smooth waveform is almost quadratic for $t \in [0, t_a]$, no iterations are necessary to find the distance $p(0, t_a)$ and the segment length t_1 (the other boundary point for the actual segment is defined for $t_0 = 0$). A typical PWL signal versus the smooth response x is presented in Fig. 2.

From electrical point of view, the PWL signals represent voltages. The blocks used for synthesis are assumed to be unidirectional, so that currents are usually not accounted for. Electrical effects that require bi-directional signal flow (e.g. in case of a transmission gate) or tight feedback loops should be avoided by means of hiding them inside the building

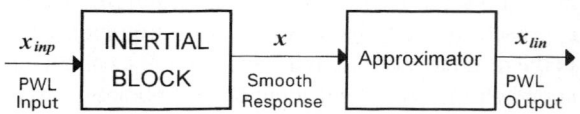

Fig. 1. Conversion of PWL signal.

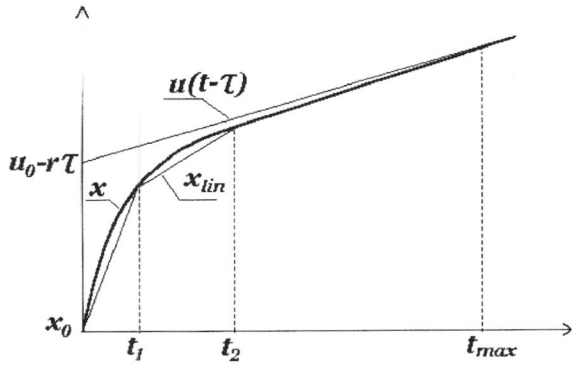

Fig. 2. PWL approximation of output signal x.

blocks. In this way, the PWL simulation proceeds usually with no iterations, and also no numerical stability problems appear regardless the step size controlled by p_{max}.

On the other hand, when feedback loops have to be modeled, the PWL algorithm may require iterations that are not supported by typical discrete simulators, like V-System [9]. Fortunately, in most cases encountered for mixed analog-digital networks, the feedback loop is controlled by an external signal, e.g. a clock that, in fact, cuts the loop. As a consequence, no iterations are required for those networks [12].

To show the adequacy of the presented approach to analog modeling at the functional level, we give an example of an amplifier [6,10,11]. Usually the following specifications must be accounted for it: the gain, the dominant pole, the saturation, the output resistance and the slew rate. For small input/output signals a fully linear macromodel is sufficient. However, to cover the possibility full range of input amplitudes the nonlinear function $f(\bullet)$ and a special slope limiting mechanism (SLM) [5] must be used. Each linear segment of the input signal u_{in} is checked for the slew rate (SR) parameter. If the segment slope $du_{in}/dt > SR/k$ and the SLM has not been activated for the previous segment, the SLM begins to act not until the fast change at the input exceeds some threshold, i.e. $\Delta u_{in} > u_{th}$, where u_{th} denotes the threshold voltage that puts the amplifier input stage into saturation. The segment slope above the threshold is then reduced to the limit SR/k (k is the amplifier gain), and the next PWL segments are shifted appropriately along the time axis to avoid time discontinuities. An algorithm for the SLM is given below in form of a pseudo code (the corresponding VHDL description is rather excessive).

```
SLM algorithm/*check for input slope*/
  if abs(du_in/dt) <= SR/k then
    SLM_out := u_in/*SLM inactive*/
    else {
      if (SLM has not been activated)
      then
        {SLM_out:=u_in for fast Δu_in<=u_th;
         SLM_out:=(u_in with reduced
           slope) for Δu_in>u_th
        }
      else SLM_out:=(u_in with reduced
        slope) for whole segment;
      shift the next segments of u_in
      /*avoid discontinuities*/
    }
```

Beside the SLM, two cascaded building blocks are required. The structure of the PWL amplifier macromodel is given in Fig. 3.

Denoting by u_{in}^* the signal obtained from SLM, for the first inertial block of the amplifier macromodel we have:

$$\tau \frac{dx}{dt} + x = ku_{in}^*, \quad \text{when } |ku_{in}^*| \leq U_{os} \quad (5)$$

$$\tau \frac{dx}{dt} + x = U_{os}, \quad \text{when } |ku_{in}^*| \leq U_{os} \quad (6)$$

where τ is the inverse of a dominant pole frequency ω_0 ($\tau = 1/\omega_0$) and U_{os} is an output saturation voltage. For the second building block we have:

$$\tau_0 \frac{dy}{dt} + y = x \quad (7)$$

where $y = u_{out}$ is the amplifier output signal and τ_0 involves the output resistance R_{out} and output capacitance C_{out} as well as a capacitive load C_l.

In Fig. 4 the functioning of the SLM is shown. The u_{in} signal consists of four PWL segments. The slopes of the first two segments exceed SR/k, but the SLM limits only the slope of the second one above the threshold voltage (that has been assumed to be equal 80 mV). It corresponds to a noninverting amplifier (with bipolar transistors) of the $SR = 0.5$ V/ms, the closed loop gain $k = 50$, and $\omega_0 = 125{,}600$ rad/s ($f_1 = 1$ MHz). The slopes of the next segments need not be limited since they are less than SR/k. The solid line denotes the original input signal and the dashed line the limited one.

In Fig. 5 the amplifier output waveforms for u_{in} are given. The u_{out}^* solid line has been obtained with the

94   J. Dąbrowski and A. Pułka

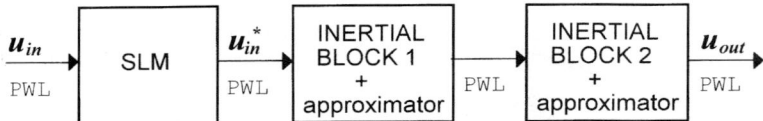

Fig. 3. Structure of amplifier macromodel.

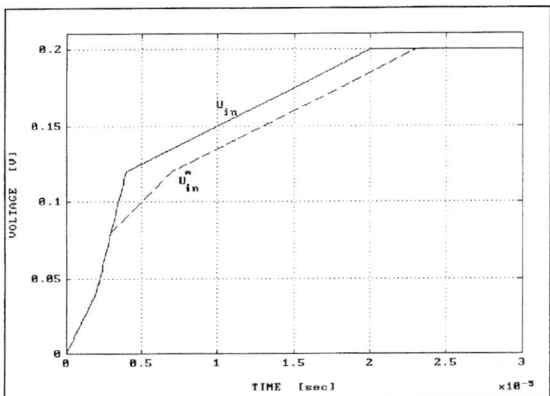

Fig. 4. Slope limiting mechanism.

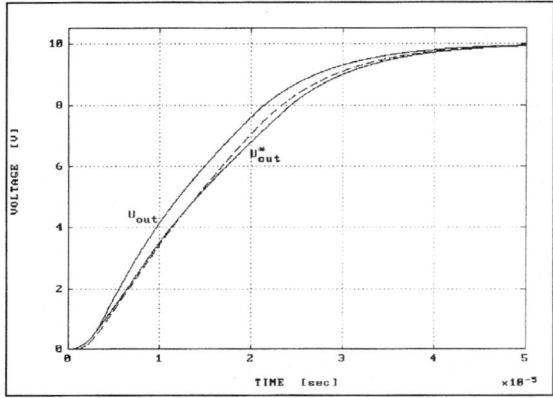

Fig. 5. Amplifier time responses.

SLM, whereas the u_{out} solid line without of it. Both curves are the original smooth responses of the inertial block defined by equations (5–7). For comparison the SPICE estimate of the u_{out}^* is also plotted (dashed line). For other values of k ($k\omega 0 =$ const.) the time responses for various input signals are also good approximation of the SPICE estimates.

Since explicit formulas are used to predict the smooth responses and their PWL approximations, the behavioral representation of the analog models seems to be the most efficient approach. Moreover, the boundary points of subsequent PWL segments may be regarded to as some discrete events. Apparently, the models of interest are better suited to a discrete, event-driven simulation technique, rather than to other methods having their origin in solution of ordinary differential equations (e.g. SPICE-like), widely used in analog simulation at lower levels of abstraction.

As mentioned earlier, unlike the traditional analog simulation, the PWL algorithm works usually with no iterations and is numerically stable for any practical value of the approximation accuracy p_{max}. Consequently, the resulting PWL segment lengths (step sizes) do not influence the algorithm stability. It is also to be pointed out that when iterations cannot be avoided (in case of large analog loops), the PWL algorithm may be supported by the waveform relaxation technique, which is stable under mild conditions [12].

3. Behavioral Description of Models

To proceed the algebraic formulas necessary for the models, an analog kernel has been derived as a separate package. It contains all necessary functions and constants. Besides, since the VHDL is a strongly-typed language [8], we had to define functions for type conversions from real- to time-type and vice versa, as well.

The interface part of our models contains parameters of the input and the output signals. Since it is a set of real- and time-typed signals, the declaration of the entity ports is not typical. Moreover, the entity involves a generic list of parameters to define the model/block (e.g. a time constant, a gain) or to define the accuracy of the PWL approximation. The interface part of our models is shown in Example 1.

Example 1: interface of intertial block

```
ENTITY Inertial_Block IS
  GENERIC (
      Pmax     :real:= 0.08;
      EPSILON6 :real:= 0.01;
      Tau      :real:= 100.0);
```

```
PORT  (U0, X0, R   : IN real;
       Tend        : IN time;
       X1, U1      : INOUT real := 0.0;
       TimeStep    : INOUT time := 0 ns;
       StartOut,
       EndOut,
       ROut        : INOUT real := 0.0;
       TendOut     : INOUT time := 0ns);
END Inertial_Block;
```

In particular, EPSILON6 denotes here the maximum value of the third order Lagrange rest (mentioned earlier) that allows to control the range of the quadratic approximation T_a. The symbol Tau stands for the time constant of the block with initial condition X0. Similarly, U0 is an initial value of the input, and R denotes its slope. Consequently, X1, U1 are to be assigned values of the output and input, respectively, after the computed time step.

The behavior of each model/block is defined by a single process with the **wait** statement at the end. We decided to construct the process with an empty sensitivity list and to activate it with the **wait for** statement. The parameter of this statement – T_1 is calculated during the execution of the previous run of the process.

It is also possible to use the sensitivity list with a strobe signal. However, in this case an additional process statement would be required and the order of execution of both the processes could influence the correctness of the model. Moreover, this solution requires some extra signals, so it is apparently more complicated.

In our approach some artificial data objects (variables) are used, which tie the model together (see Example 2).

Example 2: fragment of architecture of integrator

```
ARCHITECTURE behavior OF integrator IS
BEGIN
   PROCESS
   VARIABLE StepCounter    : integer    := -1;
   VARIABLE Initial        : boolean    := true;
   VARIABLE Start          : boolean    := true;
   VARIABLE Stop           : boolean    := false;
   VARIABLE T1,DeltaTime   : real       := 0.0;
   VARIABLE
   Xactual,Xlin0,Xsqr0     : real       := 0.0;
   VARIABLE alpha,Rt       : real       := 1.0;
   VARIABLE LinVal,SqrVal  : real       := 0.0;
   VARIABLE Tstep, T0      : time       := 0 ns;
   VARIABLE Tlastend       : time       := 1 ns;
BEGIN
   IF Initial THEN
         Xactual := Xstart;
         Xlin0 := Xstart;
         Initial := false;
   ELSE
   (...)
```

The signals from the entity list are asserted only once at the end of the process. During the execution some parameters are temporarily constant. Their changes are caused by input signals (e.g. slopes when entering a new PWL segment). Calculation of these parameters depends upon the **if** statement (see Example 3). Notice that this solution is similar to the idea of the sensitivity list.

Example 3: checking of input signals transitions for integrator

```
IF (Xend'EVENT OR Tend'EVENT) THEN
    IF NOT(Start) THEN
         Xactual : Xlin;
         Xlin0 := Xlin;
         Xsqr0 := Xsqr;
         StepCounter := 0;
    END IF;
    T0 := NOW;
    Deltatime := time_to_real
              (Tend-T0);
    alpha := (Xend - Xactual) / (2.0*
              DeltaTime);
    IF alpha = 0.0 THEN
       T1 := DeltaTime;
    ELSE
       T1 := 2.0*SQRT(Pmax/
              abs(alpha));
    END IF;
    Tstep := real_to_time(T1);
    Start := false;
END IF;
```

The latter fragment of the VHDL code involves also an algorithm for the approximator, that in case of purely quadratic output waveforms for the integrator is much simpler than for the intertial block. The remaining computations and signal assertions proceed independently from the input signal changes.

Example 4: VHDL implementation of approximator for inertial block

```
T0 := time_to_real(NOW);
 IF(Tmax > T0) THEN
    Ta:      = Tau*(EPSILON6 / abs(Xactual - Uactual +
             R*Tau))**0.33;
    Xta      := Uactual + R*(Ta - Tau) + (R*Tau + Xactual
             - Uactual)*EXP(-Ta/Tau);
    Xlta2    := (Xactual + Xta)/2.0;
    Xta2     := Uactual + R*(Ta/2.0 - Tau) + (R*Tau +
             Xactual - Uactual)*EXP(Ta/(2.0*Tau));
    Pta      := abs(Xlta2 - Xta2);
    IF Pta > Pmax THEN
            T1 := Ta * SQRT(Pmax/Pta);
    ELSE
            T1 := Ta;
    END IF;
    IF (T0 + T1 > = Tmax) THEN
        T1 := Tmax - T0;
    END IF;
    Xactual := Uactual + R*(T1 - Tau) + (R*Tau +
            Xactual - Uactual)*EXP(-T1/Tau);
    Uactual := Uactual + R*T1;
    Tstep := real_to_time(T1);
    TimeStep < = Tstep;
    X1 < = Xactual AFTER Tstep;
    U1 < = Uactual AFTER Tstep;
 END IF;
```

Here, Xta stands for $x(t_a)$, Xlta2—for $x_{lin}(t_a/2)$, Xta2—for $x(t_a/2)$, and Pta denotes $p(0, t_a)$. Clearly, T1 is the actual segment length. The last lines of the code depicted in Example 4 present concurrent signal assignments that follow the approximation part. In particular, the obtained signal values (stored as variables) Xactual and Uactual have to be assigned to the signals X1, U1 after the calculated time step Tstep. Additionally, the TimeStep output signal is used to drive the actual value of the segment length (time step). This signal is required when the considered block is followed at the output by another block.

4. Simulation Results

In this section a few results of simulation examples are presented. All of them have been obtained by means of the V-System (Model Technology VHDL Simulator) [9], which proved to be a flexible simulation tool.

In Fig. 6 the input and the output for the first order low-pass filter is given. The basic specifications of the model are as follows: the DC gain $k_0 = 1.0$ [V/V], the dominant pole $\omega_0 = 500 \cdot 10^3$ [rd/s].

The output waveform of this filter resembles the original smooth response.

The other example addresses an integrator unit with the time constant $T = 0.6\,\mu$s. The PWL waveforms obtained for this model with relatively high accuracy of $p_{max} = 0.08$ V are shown in Fig. 7. For comparison in Fig. 8 the PWL-waveforms computed for $p_{max} = 0.25$ V are given. The number of the PWL segments, the output response consists of, is in this case smaller.

In Fig. 9 a model structure of the successive approximation A/D converter is presented. It consists of three basic units: the voltage comparator, the successive approximation register (SAR) and the D/A converter. Eight-bit version of this model has been implemented. The comparator model is based on a two-stage cascade of inertial blocks. This structure is capable to mimic adequately the comparator timing specifications including the influence of initial polarisation and the overdrive at its input [11].

The D/A converter makes also use of the inertial block at the output. Its control part follows the fundamental formula: $U_{out} = \Sigma 2^i a_i \Delta U$ where $i = 0 \cdot 7$ and ΔU represents the resolution (here 16 mV). The a_i parameters are set either to 0 or to 1 with respect to the

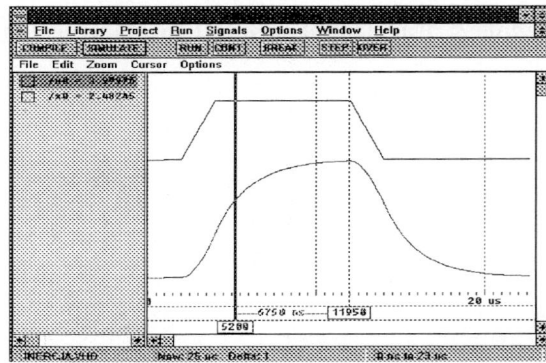

Fig. 6. Waveforms of low-pass filter ($p_{max} = 0.08$ V).

In this way, a mixed-signal simulation is faced in this example. Hence, we had to incorporate some objects to the model to convert signals from PWL to std_logic in case of *Cout* signal and from std_logic to PWL in case of the drivers that belong to the D/A.

An example of the timing waveforms obtained for the discussed A/D converter are shown in Fig. 10. In this case the simulation process depends heavily on the clock signal, which inactive simply breaks the feedback loop. As a consequence, the signal does not propagate free through the loop, so that the PWL simulation proceeds segment by segment, with no iterations (as mentioned in Section 2).

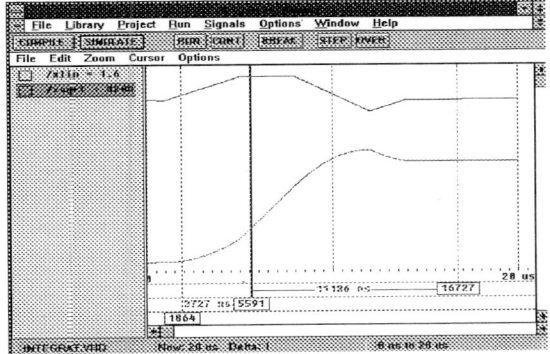

Fig. 7. Waveforms of integrator ($p_{max} = 0.08$ V).

digital input of this unit. Since this operation is physically involved with switching of signals inside the converter, any change of the a_i coefficients is controlled by their own drivers, which are modeled as digital objects with a prescribed inertial delay. In this way the D/A converter is defined as a mixed-signal model.

On the other hand, the SAR consists of digital blocks and data structures (signals and variables). The behavior of the SAR is defined at RT-Level of abstraction.

5. Summary

A discrete approach to modeling of analog units at the functional level has been presented. The method is based on PWL modeling technique and makes use of its discrete nature. The typical analog building block our synthesis is based on is accompanied by the approximator, which replaces the original smooth waveforms with PWL segments. Because the models are defined by some explicit formulas, behavioral VHDL description is used to create their architectural

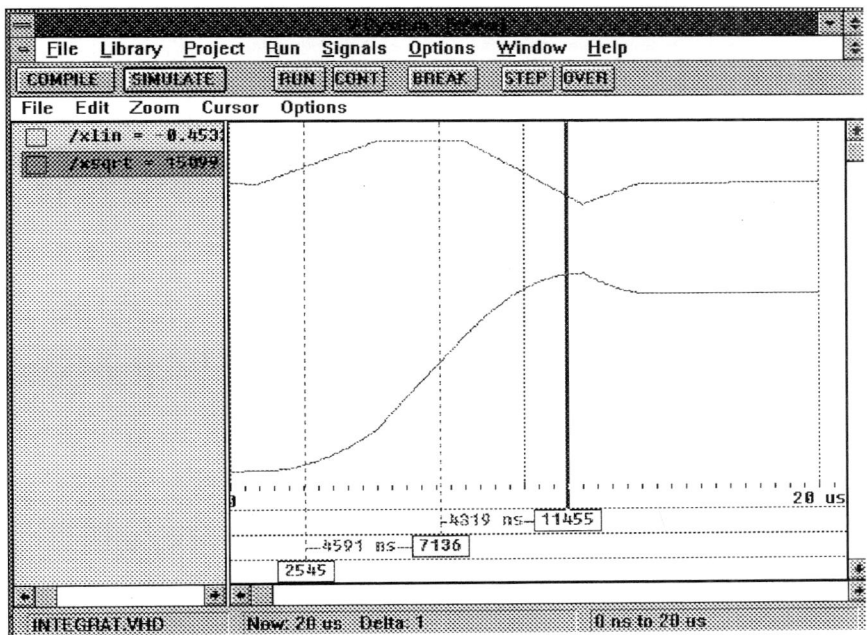

Fig. 8. Waveforms of integrator ($p_{max} = 0.25$ V).

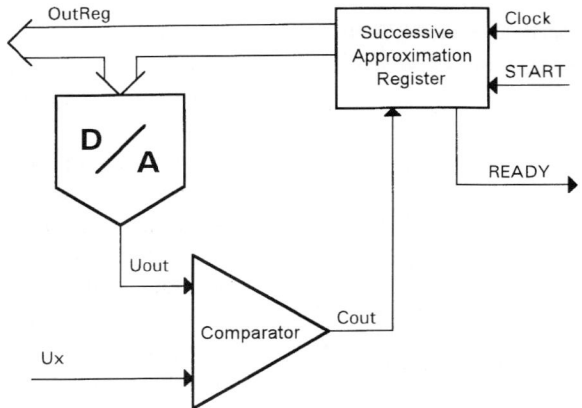

Fig. 9. Functional-level model of successive approximation A/D converter.

bodies. Since the modeling environment we employ is digital, our models are also well suited to mixed-signal A/D simulation. In this case the behavioral description of digital part of the circuit seems to be the most advantageous. Clearly, to support this kind of simulation [7] std_logic to PWL and PWL to std_logic conversions have to be defined. This approach differs from the fully unified PWL treatment of all the A/D circuit parts [5]. However, the latter is adequate only at the gate level. The representation of digital units at higher levels of abstraction is not possible for it, unlike in our approach.

References

1. J. M. Berge. *Modeling in Analog Design*. Kluwer Academic Publishers, 1995.
2. A. Mantooth. *Modeling with an Analog Hardware Description Language*. Kluwer Academic Publishers, 1994.
3. J. M. Berge, A. Fonkoua, S. Maginot, and J. Rouillard. *VHDL Designer's Reference*. Kluwer Academic Publishers, 1992.

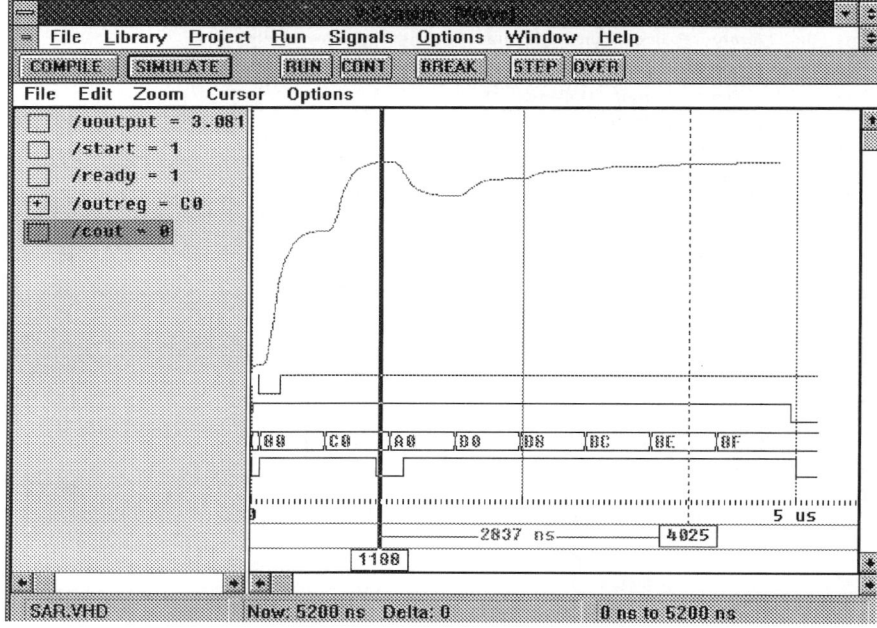

Fig. 10. Waveforms of A/D converter from Fig. 9.

4. J. P. Mermet. *Fundamentals and Standards in Hardware Description Languages.* NATO ASI Series, Kluwer Academic Publishers, 1993.
5. G. Ruan, J. Vlach, J. Barby, and A. Opal. "Analog Functional Simulator for Multilevel Systems." *IEEE Trans. on CAD* 10(5), pp. 565–575, May 1991.
6. J. Dąbrowski. "Functional-Level Analog Macromodeling with Piecewise Linear Signals." *Proceedings of EURO-DAC'95* Brighton, pp. 222–227, September 18–22, 1995.
7. R. A. Saleh and A. R. Newton. *Mixed-Mode Simulation.* Kluwer Academic Publishers, 1990.
8. IEEE Standard VHDL Language Reference Manual, IEEE Std 1076-1993, IEEE Standards.
9. V-System/Windows User's Manual, VHDL Simulation for PC's Running Windows & Windows NT, Ver. 4.2f, Model Technology Inc., USA 1994.
10. R. Baraniecki and A. Rosiski. "An Approach to the VHDL Modeling of Operational Amplifiers." *Proceedings of 2nd ATC: Mixed Design of VLSI Circuits*, Krakow, Poland, pp. 205–210, May 29–31, 1995.
11. J. Dąbrowski and J. Konopacki. Implementation of A/D Network Macromodels in PWL Functional-Level Simulator. Bull. of the Polish Academy of Sciences, *Technical Sciences*, 44(3), pp. 293–312, 1996.
12. J. Dąbrowski. Waveform Relaxation-Based PWL Simulation of Analog and Mixed A/D Networks, prepared for publication.

Jerzy Dąbrowski was born in Katowice, Poland in 1952. He obtained his M.S. (with honor) and Ph.D. degrees in Electrical Engineering from Silesian Technical University, Gliwice, Poland in 1976 and 1987, respectively. His first research interest focused on design and analysis of switched-mode power supplies. The result of that work are 14 certified patents. The current interests of Dr. Dąbrowski are in ICs macromodeling, macrosimulation, timing verification of VLSI designs based on VHDL-AMS as well as design for testability (DfT). He is an author/co-author of 40 publications merely in the area of analog and mixed A/D macromodeling, and macrosimulation. Actually, Dr. Dąbrowski is an Assistant Professor in the Institute of Electronics at Silesian Technical University, Gliwice, Poland. He is also a member of Polish Society of Electrical Engineering (PTETiS).

Andrzej Pułka was born in Katowice, Poland in 1964. He received the M.S. and Ph.D. degrees in Electrical Engineering from Silesian Technical University, Gliwice, Poland in 1988 and 1997, respectively. Currently, he is an Assistant Professor in the Institute of Electronics at Silesian Technical University, Gliwice, Poland. His research interests include: artificial intelligence techniques (especially nonmonotonic reasoning), circuits modeling in VHDL (VITAL) and VHDL-AMS, VLSI design and logic optimization.

Oversampling ΣΔ Analog-to-Digital Converters Modeling Based on VHDL

ROBERT BARANIECKI,[1] PRZEMYSŁAW DĄBROWSKI[2] AND KONRAD HEJN[2]

[1]*Institute of Electron Technology, Al. Lotników 32/46, 02-668 Warsaw, Poland*
e-mail: rbar@ite.waw.pl
[2]*Institute of Electronics Fundamentals, Warsaw University of Technology, Ul. Nowowiejska 15/19, 00-665 Warsaw, Poland*
e-mail: {przemek, K.Hejn}@ipe.pw.edu.pl

Received August 30, 1996; Accepted August 20, 1997

Abstract. The paper presents a VHDL model of an oversampling ΣΔ analog-to-digital converter created on the behavioral hierarchy level. Although VHDL has been primarily devoted to digital circuit design, it can also be applied to certain mixed-signal circuits. The model of the analog part is as simple as possible and includes only necessary parameters that enable to determine the potential resolution of a converter. The model of the digital part is described in the synthesizable subset of VHDL and parameterized according to the word length and the type of arithmetic applied. The validation process of the converter model is also shown. It is performed by a VHDL simulator and a postprocessor tool enabling to carry out FFT. Simulation results enclosed prove the efficiency of the design approach presented.

Key Words: Sigma-delta modulator, decimator, VHDL, behavioral modeling and simulation, RTL synthesis

1. Introduction

There are two basic reasons for working out behavioral models of mixed-signal circuits. The first one is their high complexity. For example an oversampling sigma-delta analog-to-digital converter (ΣΔ ADC) consists of an analog-digital ΣΔ modulator and a noninvariant digital filter called decimator [1]. The exhausted simulation of such a mixed-signal circuit is very CPU intensive especially if we try to apply a circuit level simulator like SPICE. Moreover the transistor models of mixed-signal blocks are not available at the beginning stages of the design process. The second reason is related to the top-down design methodology that recommends the validation of the model at each hierarchy level of a design process. So an efficient solution seems to be using behavioral (discrete in time) analog models. By means of them a designer can quickly validate any model of the system under consideration.

Unfortunately the appropriate tools to do this are still not available. SABER from Analogy or ELDO form Anacad is more tailored to analog than digital circuits and in the meantime VHDL-AMS is still under development. So we have applied the S̲IGNAL P̲ROCESSING W̲ORKSYSTEM (SPW) and an event-driven simulator SYNOPSYS to behavioral modeling and simulation of an oversampling ΣΔ ADC. The IEEE Std.1076 of VHDL has been the interface between them.

The primary model of an oversampling ΣΔ ADC was created in SPW environment. The VHDL code automatically obtained did not include any physical constrains such as asymmetry of thresholder levels, differences in the rise or fall times of the signals generated, etc. These imperfections have been added by hand and then the updated model has been simulated in SYNOPSYS environment.

2. A Model of an Oversampling ΣΔ ADC

A behavioral model of the oversampling ΣΔ ADC under consideration consists of two cascaded parts:
a) a 3rd order ΣΔ modulator in MASH[1] configuration [2]—mostly analog, and
b) a five-stage decimator [3]—pure digital.

The first one is shown in Fig. 1. The 1st order ΣΔ modulator is its main block, Fig. 1b. It includes a differential integrator, a one-bit d/a converter (clipper), a one-bit a/d converter (thresholder) and, additionally, an analog subtractor that creates the

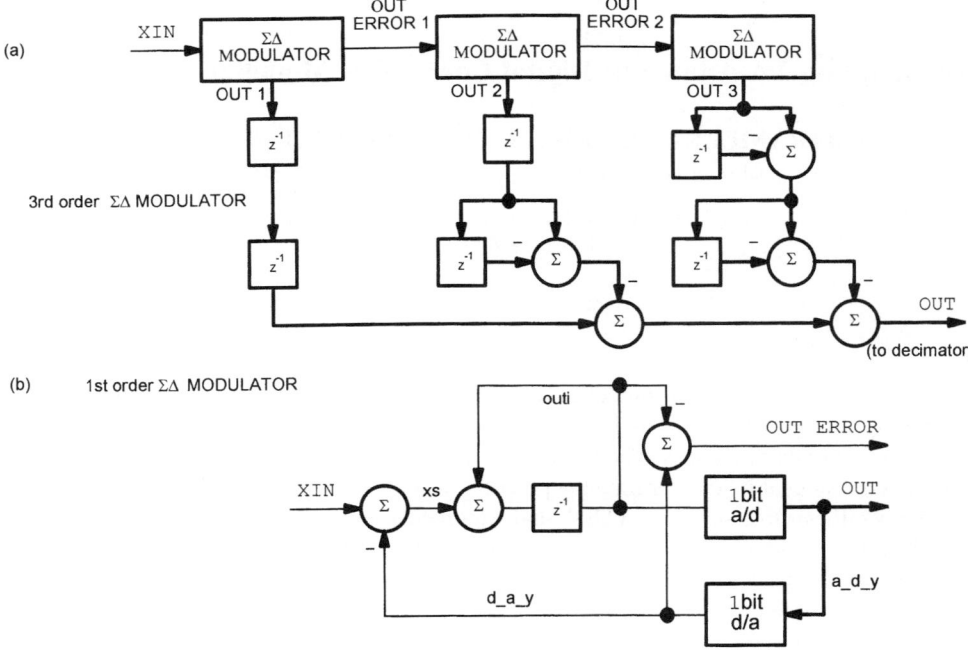

Fig. 1. A hierarchical structure of the ΣΔ modulator applied: a) 3rd order ΣΔ modulator in MASH configuration, b) 1st order ΣΔ modulator with the additional output OUT ERROR.

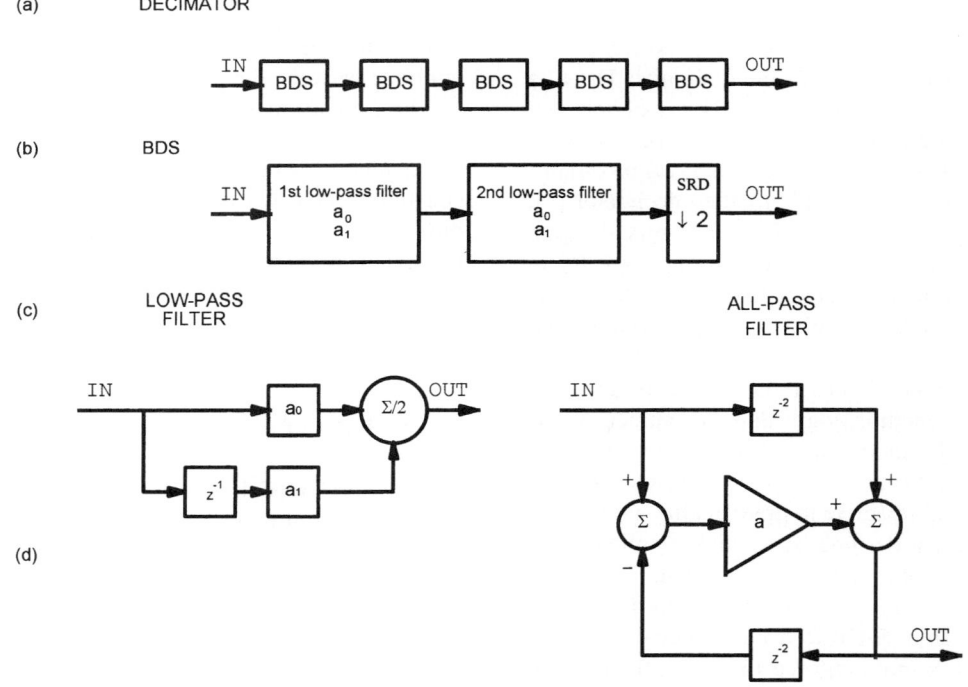

Fig. 2. A hierarchical structure of a decimator applied: a) up to five cascade of basic denominator stages (BDS), b) details of BDS consisting of two low-pass filters (LPF) and sampling ratio decreaser (SRD), c) details of LPF consisting of 2nd order all-pass filters (APF) in two-phase structure, d) details of APF with binary scaled coefficients: $a = a_0 = 1/8$ or $a = a_1 = 9/16$.

```vhdl
entity First_Order_Sigma_Delta_Modulator is
    generic (MIN_VAL, MAX_VAL              :real;  --range of input signal
             TIME_CONSTANT                 :real;  --time constant of integrator
             RIGH_LEVEL_OF_THRESHOLDER     :real;  --right threshold level
             LEFT_LEVEL_OF_THRESHOLDER     :real;  --left threshold level
             COMP_RISE_DELAY               :time;  --propagation delay of 1-bit
                                                   --a/d when output signal is rising
             COMP_FALL_DELAY               :time;  --propagation delay of 1-bit a/d when
                                                   --output signal is falling
             HIGH_LEVEL_OF_CLIPPER         :real;  --upper level of 1-bit d/a converter
             LOW_LEVEL_OF_CLIPPER          :real;  --bottom level of 1-bit d/a converter
             C_SLEW_RATE_PLUS              :real;  --slew rate of 1-bit d/a converter when
                                                   --output signal is rising
             C_SLEW_RATE_MIN               :real); --slew rate of 1-bit d/a converter when
                                                   --output signal is falling
end First_Order_Sigma_Delta_Modulator;

architecture Beh of First_Order_Sigma_Delta_Modulator is

signal xin   : real range MIN_VAL to MAX_VAL : = 0.0; --input samples
signal a_d_y_: std_ulogic : = '0';                    --output of 1-bit a/d converter
                                                      (thresholder)
signal d_a_y : real range HIGH_LEVEL_OF_CLIPPER to LOW_LEVEL_OF_CLIPPER : = 0.0;
                                                      --output of 1-bit d/a converter
                                                      --(clipper)
signal xs    : real range 2*MIN_VAL to 2*MAX_VAL : = 0.0; --subtraction node
signal outi  : real range 2*MIN_VAL to 2*MAX_VAL : = 0.0; --integrator output
begin
    xs < = xin - d_a_y;
    integrator (xs, outi, TIME_CONSTANT);
    thresholder (outi, a_d_y, HIGH_LEVEL_OF_THRESHOLDER, LOW_LEVEL_OF_THRESHOLDER,
                 COMP_RISE_DELAY, COMP_FALL_DELAY);
    clipper (a_d_y, d_a_y, HIGH_LEVEL_CLIPPER, LOW_LEVEL_CLIPPER,
             C_ SLEW_RATE_PLUS, C_ SLEW_RATE_MIN);
end Beh;
```

Sidebar 1. The VHDL code of the 1st order ΣΔ modulator

quantization error OUT_ERROR needed in the next stage. The sampling rate, due to oversampling operation, is significantly greater than the Nyquist's frequency. It is worth noting that the modulator output OUT is a binary signal. The two-level analog signal d_a_y traces on average the analog input XIN. The 3rd order MASH configuration (Fig. 1a) gives much better noise-shaping effect than the 1st order ΣΔ modulator and eliminates completely stability problems [2].

Let us focus on a VHDL model of the 1st order ΣΔ modulator, Sidebar 1. It is purely discrete in time but continuous in amplitude as it uses floating-point arithmetic for both the input and output signals. Although the model is very simple it is efficient enough to validate the properties of the oversampling ΣΔ ADC in question. It can be treated as a pattern source in the design process of an appropriate decimator architecture. The imperfect components of the modulator are described as procedures including the following parameters:
— time constant of integrator,
— hysteresis and delay time for 1-bit a/d (thresholder),
— reference levels and slew rates for 1-bit d/a (clipper).

The details of the procedures mentioned can be found in Sidebar 2.

Fully hierarchical structure of the five-stage decimator is shown in Fig. 2a. Each basic decimation stage (BDS) filters out the shaped quantization noise and simultaneously decreases the sampling rate by two, Fig. 2b [3]. BDS is implemented as a cascade of two low-pass filters (LPFs) in two-phase parallel structure with sampling rate decreaser at the output [4], Fig. 2c. For the specific coefficients $a_0 = 1/8$ and $a_1 = 9/16$, the single LPF has an excellent flat amplitude response in the pass-band but simulta-

```vhdl
procedure integrator    (signal xin   : in real;      -- input signal
                         signal xout  : inout real;   -- output signal
                         time_const   : real          -- time constant
                        ) is
    variable time_last           : time: = 0 ns;  -- instant of the last change of
                                                  -- the input signal
    variable time_delta          : time: = 0 ns;  -- temporary value of the sampling period
begin
    loop
        wait on xin;
        time_delta : = now - time_last;
        time_last : = now;
        xout < = (1.0/time_const)*xin*time_to_real(time_delta) + xout;
    end loop;
end integrator;
procedure thresholder (signal xin : in real;        -- analog input signal
                       signal xout: out std_ulogic; -- digital output signal
                       lev1, lev2 : real;           -- reference levels
                       td01, td10 : time            -- delay times
                      ) is
begin
    if rising (xin) then
        if xin > = lev1 then
            xout < = '1' after td01;
        else
            xout < = '0' after td10;
        end if;
    elsif falling(xin) then
        if xin < = lev2 then
            xout < = '0' after td10;
        else
            xout < = '1' after td01;
        end if;
    else   null;
    end if;
end thresholder;
procedure clipper(signal xin: in std_ulogic;            -- digital input signal
                  signal xout: inout real;              -- analog output signal
                  sr_p, sr_m : real) is                 -- slew rate values
    variable    t_wait       : time;                    -- time step
    variable    t_delta      : real;                    -- real step
    variable    sr           : real;                    -- temporary value of slew rate
begin
    while xout /= std_ulogic_to_real(xin) loop
        if rising(xin) then
                sr  := sr_p;
        elsif falling(xin) then
                sr  := sr_m;
        else exit;
        end if;
        t_delta:= abs((std_ulogic_to_real(xin) - xout)/sr);
        if t_delta < time_to_real(ANALOG_TIME_STEP) then
            xout < = std_ulogic_to_real(xin);
            t_wait := real_to_time(t_delta);
        else
            xout < = xout + sr*time_to_real(ANALOG_TIME_STEP);
            t_wait := ANALOG_TIME_STEP;
        end if;
        wait on xin for t_wait;
    end loop;
end clipper;
```

Sidebar 2. The VHDL code of components of the 1st order $\Sigma\Delta$ modulator

neously the unsatisfactory attenuation in the stopband. That is why the two LPFs have been cascaded. Note that the binary scaled coefficients of all-pass sections allow to replace the multiplying by bit shifting in any hardware implementation.

The basic building block of the LPF and the more so of the decimator is the second order IIR (infinitive impulse response) all-pass filter (APF). Its transfer function in Z-transform domain is following

$$H(z) = \frac{a + z^{-2}}{1 + az^{-2}}$$

It can be implemented in different ways. One of them is shown in Fig. 2d [3].

The architecture of the decimator and its model have been worked out by means of a special tool for DSP circuits. Then VHDL code was parameterized concerning the word size and type of arithmetic applied. The result for the coefficient $a = a_1 = 9/16$ is shown in Sidebar 3.

All arithmetical operations performed by APFs are fixed-point. The vector signals are expressed in the two's complement format represented as *TC* type in the VHDL code. There is a trade-off in the vectors' dimension and the type of arithmetic applied. Formats of port vector signals in the APF VHDL code are expressed as "1 downto WIDTH2". The WIDTH stands for a vector size. The left boundary represents the number of bits of the integer part. Such format is required by the arithmetical procedures.

The fxpAshift, fxpAdd, fxpSub round procedures are imported from the COMDISCO arithmetic package available inside the SYNOPSYS library. The constant *ROUNDING* characterizes modes of the precision loss. There are a few types of loss of the precision modes in the COMDISCO package: round to minus infinity for truncation, round to plus infinity, round to zero for magnitude truncation, round for typical mathematical round, and convergent round to nearest even number. The constant *ROUNDING* characterizes the selected mode of the precision loss. The smallest ripples of the amplitude response are obtained by convergent rounding.

Let us focus on the multiplication with coefficient 9/16, see Fig. 3. This coefficient is binary scaled and the multiplication can be implemented as a shifting. It is easy to notice that 9/16 is the sum of 1/2 and 1/16. Thus the multiplications by 1/2 or 1/16 have been realized as right shifting by 1 and 4 respectively. To keep the precision of computations the boundaries of vectors carrying the multiplication products are different from the boundaries of *S_IN* and *S_OUT*. The variable *VMULT12* has format "0 downto WIDTH-3". In case of the variable *VMULT116*, its format is "-2 downto WIDTH-5". Additionally, the final multiplication result *S_MULT* is a vector that is wider by one bit than others. It was done to remove oscillations appearing in the APF with coefficient 9/16.

3. The Validation Process of the Converter Model

The validation process, see Fig. 4, of the converter model is based on two tools: a VHDL simulator and a waveform calculator. At first, the converter model is simulated using a VHDL simulator of SYNOPSYS. A pure sine wave is applied as a stimulus. The results obtained concern only the time domain. To learn the converter behavior in the frequency domain, a VHDL simulation output file is imported into a waveform calculator inside SPW system to carry out spectral analysis. Some results of the spectral analysis are shown in Fig. 5. Note that the structure analyzed has approximate potential of an 18-bit ADC.

Table 1 presents some experimental results corresponding to the simulations that have been carried out on the SUN4 workstation. Simulation time depends exponentially on the $\Sigma\Delta$ modulator order as well as on the number of decimator stages.

4. Conclusions and Recommendations

The IEEE Std. 1076 of VHDL has turned out to be a powerful tool for oversampling $\Sigma\Delta$ ADC modeling and simulation. Such an approach is appropriate for validation of the design if an analog part of a mixed signal circuit is relatively simple. The architecture validation and the functional testing of mixed-signal circuits will become much easier if VHDL description is applied at the early stages of the design process.

According to the results shown in Table 1, the main disadvantage of the method presented is its relatively long simulation time. A Native-Compiled-Code based simulator can be a right solution speeding up the simulation.

The future work will concern further developing of the models already proposed. The VHDL-A (IEEE

```vhdl
entity APF_916 is
    generic(WIDTH :integer;                         - - decimator word size
                                                    - - rounding type
            ROUNDING :BIT_VECTOR(15 downto 0) := FXP_ROUND_CONVERGENT;
            );
    port(res : in std_logic;                        - - reset
         clk : in std_logic;                        - - clock
         s_in: in tc(1 downto -1*(WIDTH-2));        - - input (operand)
         s_out: out tc(1 downto -1*(WIDTH-2)));     - - output (result)
end APF_916;
architecture Beh of APF_916 is
  signal s_sub   : tc(s_in'high downto s_in'low);     - - subtraction result
  signal s_sum   : tc(s_in'high downto s_in'low);     - - summing result
  signal s_mult  : tc(s_in'high downto (s_in'low-1));- - multiplication result
  signal s_delay1: tc(s_in'high downto s_in'low);     - - branch
  signal s_delay2: tc(s_in'high downto s_in'low);     - - delays

begin
- - Multiplication by 1/8 = right shifting by 3
- - Multiplication by 9/16 (1/2 + 1/16) == right shifting by 1 and 4

  SHIFTING :process(stw,xsub)

  variable v_mult12:tc((s_in'high-1) downto (s_in'low-1));  - -result of multiplication by 1/2
  variable v_mult116:tc((s_in'high-3) downto (s_in'low-3));- -result of multiplication by 1/16
  variable v_mult :tc(s_in'high downto (s_in'low-1));       - -final multiplication result
  begin
      fxpAshift(ROUNDING,s_sub,-1,v_mult12);
      fxpAshift(ROUNDING,s_sub,-4,v_mult116);
      fxpAdd(ROUNDING,v_mult12,v_mult116,v_mult);
      s_mult <= v_mult;
  end process SHIFTING;
DELAY :process(res,clk)
  variable v_delay1, v_delay2 : tc(s_in'high downto s_in'low); - -branch delays
  begin
      if res = '1' then
          s_delay1 <= (others => '0');
          s_delay2 <= (others => '0');
          v_delay1 := (others => '0');
          v_delay2 := (others => '0');
      elsif clk = '1' and clk'event then
          s_delay1 <= v_delay1;
          s_delay2 <= v_delay2;
          v_delay1 := s_sub;
          v_delay2 := s_sum;
      end if;
  end DELAY;
  fxpSub(ROUNDING,s_in,s_delay1,s_sub);
  fxpAdd(ROUNDING,s_mult,s_delay2,s_sum);
  s_out <= s_sum;

end Beh;
```

Sidebar 3. The VHDL code of APF with coefficient 9/16.

Oversampling 107

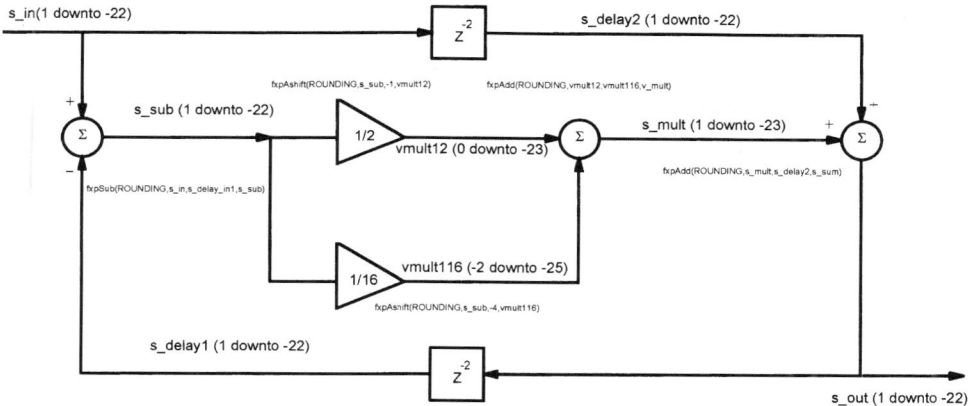

Fig. 3. The scheme of arithmetic operations executed in 24-bit APF with coefficient 9/16.

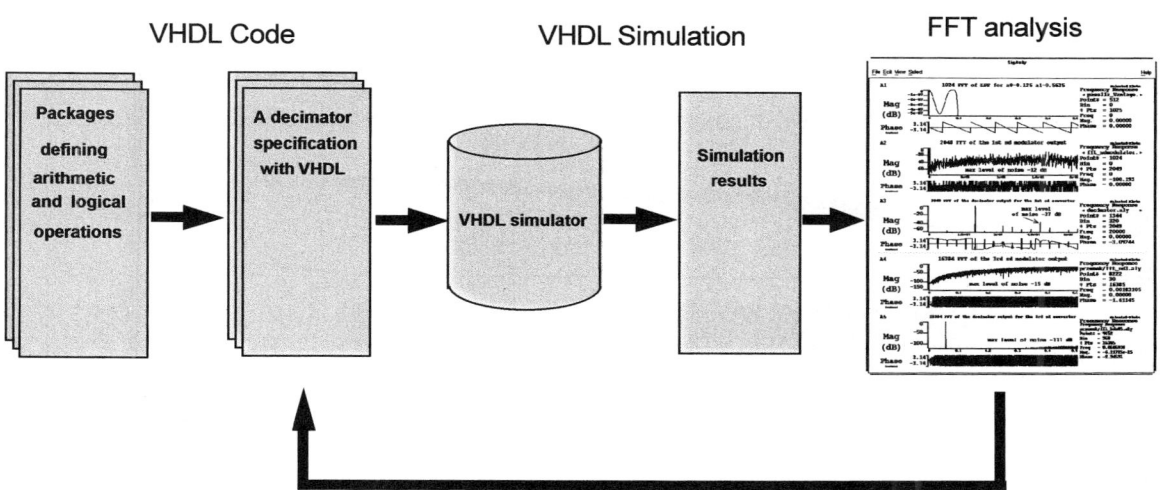

Fig. 4. A diagram of the decimation filter validation process.

Table 1. The experimental simulation results of the oversampling ΣΔ ADC.

	The 1st Order ΣΔ Modulator	Decimator (three stages)	The 3rd Order ΣΔ Modulator	Decimator (five stages)
Number of samples	65 536	65 536	640 000	640 000
Number of events	992 458	734 215	9 893 754	7 125 131
CPU time	7 min.	4.5 min.	67 min.	58 min.

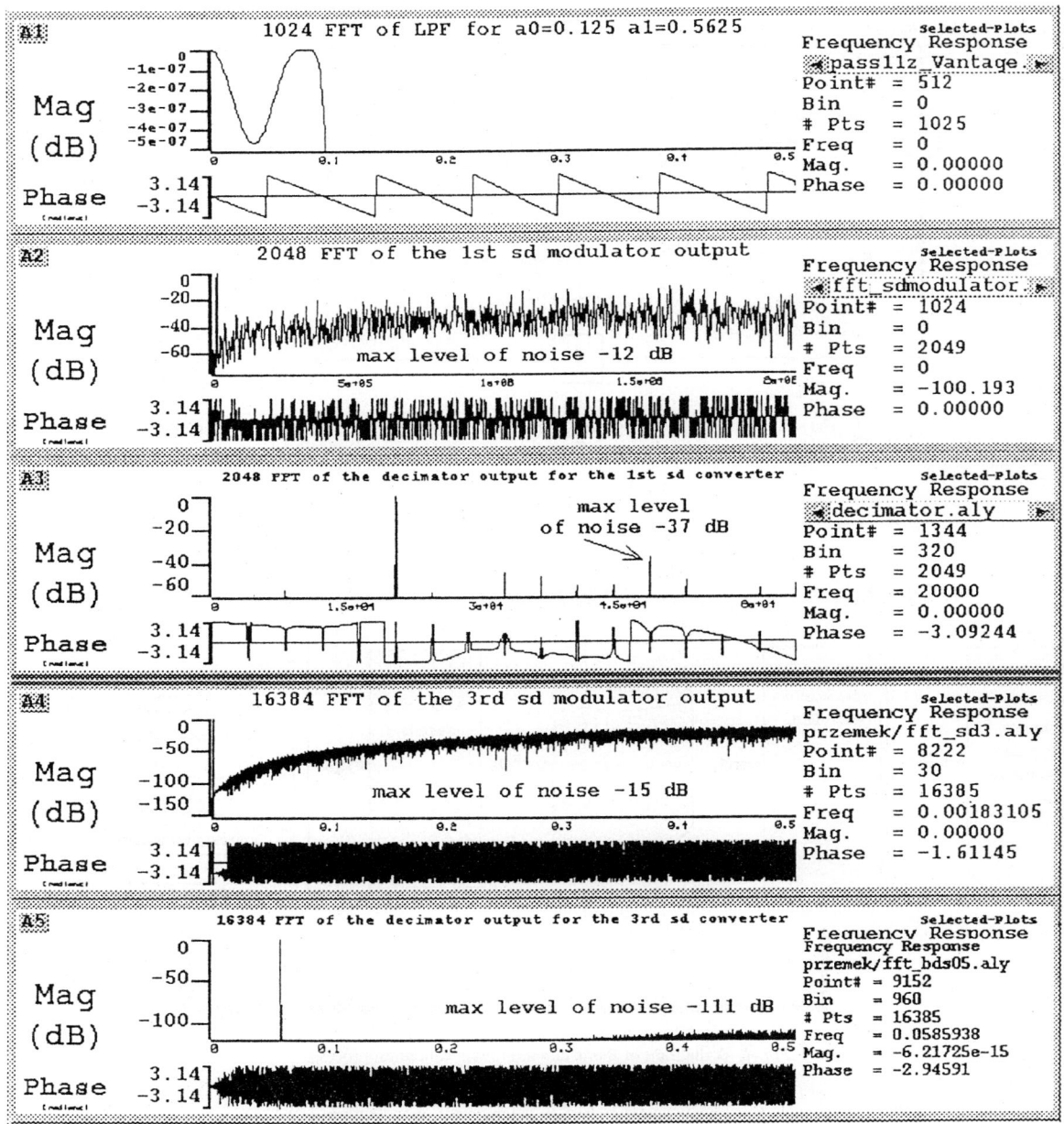

Fig. 5. The analysis in frequency domain.

A1—the amplitude response of the LPF (Fig. 3), ripples below of 0.5 μdB;

A2—the spectrum of the 1st order $\Sigma\Delta$ modulator (Fig. 1b) output for a harmonic input of fnorm = 10/2048 (note the spike near the vertical axis and the shape of the noise), SNR = 12dB;

A3—the spectrum at the decimator (Fig. 4) output for the 1st $\Sigma\Delta$ modulator at its input, SNR = 37dB;

A4—the spectrum of the 3rd order $\Sigma\Delta$ modulator (Fig. 1a) output for a harmonic input of fnorm = 7500/4096000 (note that the noise is more shaped than for the 1st order $\Sigma\Delta$ modulator (Fig. 7.A2)), SNR = 15dB;

A5—the spectrum at the decimator (Fig. 4) output for the 3rd order $\Sigma\Delta$ modulator at its input, SNR = 111dB.

P1076.1) will be applied soon and some comparison investigations will be carried into effect. The VHDL synthesizable core for the decimator will be also an object of the forthcoming research.

Note

1. MASH—multi-stage noise shaping modulation technique that avoids limit cycles and stability problem inherent for 1st order $\Sigma\Delta$ modulator [2].

References

1. J. C. Candy and G. C. Temes. "Oversampling Delta-Sigma Data Converters." IEEE Press, 1992.
2. W. Chou, P. W. Wong, and R. H. Gray. "Multistage sigma-delta modulation." *IEEE Trans. Inform. Theory* 35(4), pp. 784–796, July 1989.
3. K. Hejn, P. Murphy, and I. Kale. "Measurement and Enhancement of Multistage Sigma-Delta Modulators." IEEE Instrumentation and Measurement Technology Conference, New York, pp. 545–551, 12–14 May 1992.
4. R. A. Valenzuela and A. G. Constantinides. "Digital signal processing schemes for efficient interpolation and decimation." *IEE Proceedings* 130, Pt. G, No. 6, pp. 225–235, December 1983.

Przemyslaw Dąbrowski was born in Bialystok, Poland in 1969. He received his M.S. degree in Electronics from Warsaw University of Technology in 1994 and M.Sc. in DSP from University of Westminster, London in 1995. Currently he is Ph.D. student at Warsaw University of Technology.

Robert Baraniecki was born in Warsaw in 1968. He received his M.Sc. degree in Microelectronics from the Warsaw Technical University in 1993. After his degree he joined the VHDL research group at the Institute of Electron Technology in Warsaw. Then he began his work towards a Ph.D. in VHDL modeling of analog and mixed-signal circuits. He is the author of more than ten papers related to this topic. In 1995 he received a prestige award from the Foundation for Polish Science.

Konrad Hejn was born in Warsaw, Poland in 1943. He received the masters degree in control techniques, in 1966, and the Ph.D. degree in electronics in 1976, both from Warsaw University of Technology. Upon graduation in 1966, he joined the Warsaw University of Technology, and has since been working in the field of computer-aided measurement and instrumentation. In 1991 and 1993 he was visiting professor at University of Westminster, engaged in work related to DSP. Currently his primary interests are in oversampling delta-sigma data converters. He presently serves as a head of Measurement System Group in the Institute of Electronics Fundamentals.

VHDL Modeling of Optoelectronic Interconnect Networks

SEUNGUG KOH

University of Dayton, Department of Electrical and Computer Engineering, 300 College Park, Kettering Lab 251, Dayton, OH 45469-0226

Received October 4, 1995; Accepted January 20, 1997

Abstract. High performance computer and communication systems of today demands improvements in communication delay and connectivity. Optoelectronic interconnect is well recognized as a technology to provide the necessary improvement. However the complexity of optoelectronic systems is motivating a need for CAD tools which can simulate electronic and optoelectronic device and circuit behaviors simultaneously and at multiple levels of abstraction. This paper describes VHDL modeling of optical signals and integrated optical waveguides which enables seamless integration of mixed electronic/optoelectronic simulations in an established electronic design automation environment.

Key Words: VHLD modeling, optoelectronic interconnection, system clocking

1. Introduction

Advances of semiconductor fabrication process made it possible to design and fabricate chips with several millions of transistors operating at very high speed (i.e. over several hundred MHz) [1]. And these advances are dynamically incorporated with innovative hardware organizations of modern integrated circuits (ICs) and achieved remarkably high performance at low cost. However there are also some negative architectural implications such as: a) communication is more expensive than computation because high-performance Si or GaAs integrated circuits require increasingly high-speed signal interconnection among their processing units and parallel/interconnection-intensive hardware organization demands longer interconnection distance and larger signal interconnection area, b) chip boundaries have the effects of limiting the input/output (I/O) data transmission bandwidth and of creating a substantial disparity between on-chip and off-chip communication delays [2]. Therefore designing faster and larger ICs for high-performance computer and communication systems demands improvement in communication delays and connectivity.

Optoelectronic interconnection is well recognized as a technology to alleviate the above communication bottleneck of high-performance computer systems and communication networks [3]. Optoelectronic interconnection has known advantages such as large information capacity, no electromagnetic wave interference, high interconnection density, low power consumption, high speed, and planar signal crossing. However, despite the numerous potential benefits, optoelectronic interconnect technology need to be carefully applied to the existing information processing systems such that computer or communication system designers do not need to radically modify their electronic hardware or system software. Since the possibility of the silicon-based electronic information processing technology is not fully exploited yet to reveal a fundamental performance bound, high performance information processing systems of today attempt to adopt a hybrid system which employs electronic devices for information processing purposes while applying optoelectronic devices to a critical communication network [4]. However there is no system level simulation methodology that can design such a hybrid optoelectronic information processing system. But there currently exists rich sets of electronic design automation (EDA) tools for system level modeling, particularly VHDL based EDA tools. From these observations, this paper presents a *optoelectronic interconnect simulation methodology* which can seamlessly integrate optoelectronic device simulation into the established VHDL environment. This paper specifically addresses issues of ''Guidedwave Optoelectronic Interconnect Simulation'' using VHDL to support mixed optoelectronic/electronic simulation. The use of VHDL will enable efficient system level modeling and design of optoelectronic interconnect networks by taking advantage of the existing electronic VHDL design environment without any modification. In the following section, modeling issues regarding optical signals and optoeletronic devices are addressed under the VHDL environment. Section 3 describes optoelectronic clock distribution network on MCMs with an example system level simulation that utilizes the models developed in Section 2. Finally conclusions with future development plans are summarized in Section 4.

2. Integrated Optical Waveguide Modeling using VHDL

2.1. Optoelectronic Simulation Package

Optoelectronic Simulation Package is a VHDL package describing new optoelectronic physical types useful for optical signal modeling, various integrated optical waveguide entities and architectures, and foreign import models for numerical calculations by using external C functions. A user-defined signal type *Optical* is created inside the *Optoelectronic Simulation Package* to model optical signal propagation/distribution behaviors along the waveguide networks. Like electrical potential (*Voltage*) for electronic signals, electromagnetic field intensity is used to represent signal states of optical signals. The signal type *Optical* has a user-defined VHDL record type, consisting of field distribution vector or signal intensity, wavelength, spectral width, signal rise time (i.e. signal bandwidth), polarization, and guided control bit. For high level behavioral system modeling of optoelectronic interconnect networks, a scalar signal intensity could be adequately used to represent the signal state instead of using a bulky electromagnetic field distribution vector. Signal wavelength, spectral widths, polarization and bandwidth are intrinsic characteristics of optical beams which carry signal informations. The guided control bit is a convention to represent whether an optical signal is propagating through a free space or along guided mediums like optical fibers or waveguides. Above signal attributes are interacting with optical waveguide parameters, such as dispersion, waveguide index, waveguide length, and signal losses due to coupling or beam propagation, to determine the optical signal propagation/distribution behaviors along the interconnection networks.

To facilitate the use of *Optical* signals, which have various physical attributes as above, several new physical types are exclusively defined and used for this purpose. These are *length*, *speed*, *propagation loss*, *dB loss*, and *dispersion*. These newly defined physical types with accompanying overloaded operators are extensively used for VHDL modeling of optical signals and waveguides. The overloaded operators handle various physical type conversions inside VHDL architectural descriptions of integrated optical waveguides. Several C functions are also defined as foreign import models to model various waveguide behaviors numerically.

2.2. VHDL Modeling of Integrated Optical Waveguides

There exist several basic structural configurations of integrated optical waveguides, as shown in Fig. 1, by which various guided optoelectronic interconnection networks can be constructed. Five basic waveguide VHDL entities out of these basic structural configurations are currently developed and successfully used as fundamental building blocks for complex interconnection structures like global clock distribution networks on multichip module. The five waveguide entities are *straight waveguide*, *beam splitter*, *waveguide bending*, *input coupler*, and *output coupler*.

Behavioral descriptions of these entities are written in VHDL entity-architecture combinations. Behavior descriptions of basic entities are primarily based on a coupled mode theory which is widely accepted as a reasonable first-order model for integrated optical waveguides [5–7]. In VHDL each entity may support multiple architectures employing different physical descriptions of entities. For example, it is possible to model a beam splitter, which is a basic VHDL entity, as a directional coupler or a Y branch. Both directional coupler and Y branch show identical functional behaviors and are indistinguishable if we observe only their behaviors.

Since VHDL allows multiple levels of hierarchy and multiple architectures to describe a single design, it is also possible to devise a more complex but much detailed optical waveguide behavioral or structural models as needed. For instance, an architecture for

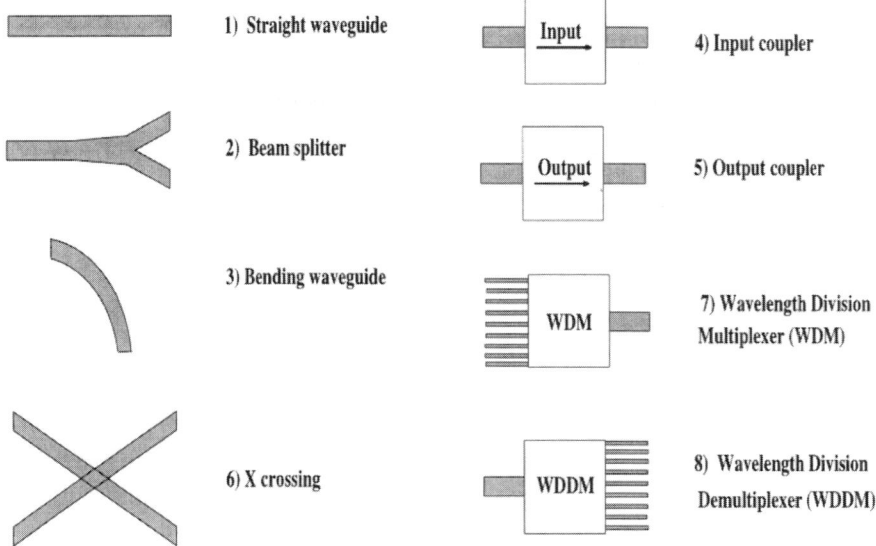

Fig. 1. Basic waveguide entities for VHDL modeling.

straight waveguide entity can be either simple behavioral relationships of input/output signals or detailed electromagnetic field distributions of a guided mode signal along the straight waveguide. Electromagnetic field distribution is sensitive to waveguide boundary conditions and needs to use a coupled-mode theory or a numerical method like a beam propagation method [5–7]. In general simpler models are preferred for high level simulations. But detailed electromagnetic field distribution of a guided mode signal is required to design a specific waveguide boundary condition. The latter utilizes low level device specifications such as total waveguide area, excess loss margin, operation wavelength, dispersion, and spectral width.

The following code is an example VHDL behavior model of an optical waveguide bending which has one input ($Op1$) and one output port ($Op2$). Fig. 2 illustrates interactions of input optical signal with physical attributes of waveguide bending entity to generate an appropriate output signal. In this model physical waveguide parameters for a bending structure are passed to the entity using a generic map. These physical parameters are waveguide crosssectional dimension, bending radius, signal propagation loss, excess loss (bending loss), material dispersion, waveguide dispersion, core refractive index, buffer refractive index, effective refractive index, and noise index. These physical parameters are incorporated with input signal ($Op1$) attributes such as wavelength, spectral width, polarization, and guided control bit and calculates an accurate output signal ($Op2$). In this example waveguide bending loss is calculated using a model by Lee which is defined inside a foreign C function *bending()* [5]. The foreign C function *bending()* assumes that a substrate coupling loss is negligible by having a sufficiently thick buffer layer between waveguide core and substrate.

3. Simulation of Synchronous Optical Clock Distribution Networks on MCM

This section describes a VHDL simulation of 16 node synchronous optoelectronic clock distribution networks (OCDN) on MCMs. A complete structural description of 16 node OCDN on MCMs is constructed and simulated by using the five basic optical waveguide VHDL entities described in Section 2 (see Fig. 1).

3.1. Synchronous Optical Clock Distribution Network on MCMs

Strategies and detailed ideas for distributing global signals on MCMs are described in [4] and [8] which

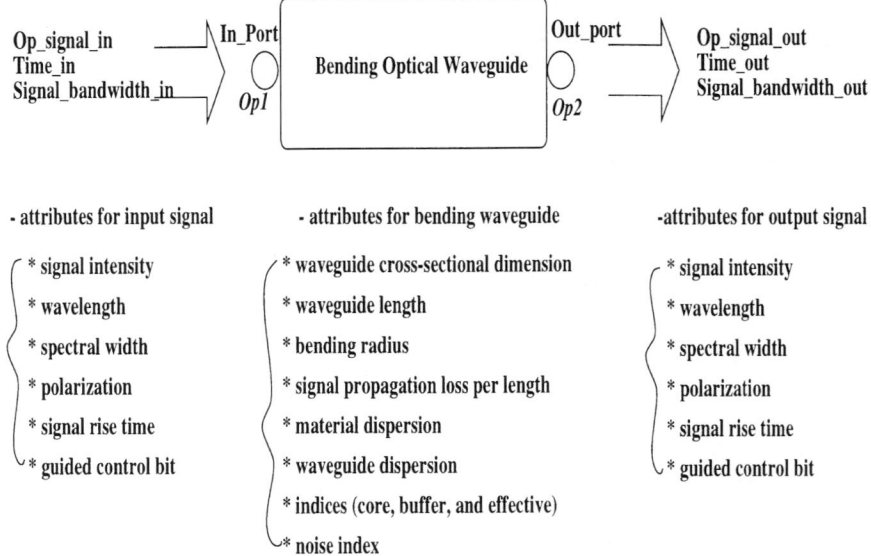

Fig. 2. Basic waveguide entities for VHDL modeling.

tried to combine advantages of MCM and optoelectronic interconnection technologies efficiently at the module level. Both MCM and optoelectronic interconnection technologies can alleviate the communication bottleneck of high-performance computer systems and communication networks. MCM contains several bare chips (dies), which are interconnected using metal interconnection layers on a substrate. MCM provides small chip-to-chip interconnection delay, small module size, and small power consumption by reducing one packaging level (i.e. individual chip packaging) and realizing miniaturalization. OCDN on MCMs are integrating MCM and optoelectronic interconnect technologies to support the operation of a large number of ICs on MCMs at very high-speed (above 500 MHz). OCDN is also utilizing a H-tree network structure which is a symmetric equidistance network for a zero or very low skew clock distribution [9] as shown in Fig. 3. Using this scheme system clock signals are equally delayed at each of the multiple destinations on a MCM for low skew clocking. Optical implementation of H-tree network on MCM can support very high-speed operation of general digital systems and improves system bandwidth, fanout, and power consumption.

3.2 Simulation Result of OCDN

We investigate behaviors of 16 node H-tree clock distribution network using a VHDL structural description. A test structure of OCDN is shown in Fig. 3 and Table 1 describes physical parameters of OCDN used for this simulation.

A total of sixty-two basic VHDL waveguide entities are used to describe a structural architecture of 16 node H-tree network and these are fifteen straight waveguides, thirty bending waveguides, fifteen beam splitters, one input coupler, and one output coupler.

During the mixed signal optoelectronic interconnect simulation for OCDN, three basic simulation modules are used as shown in Fig. 4. These modules are: a) SPICE module for laser-diode transmitters and photo-diode receivers, b) C module numerical modeling, and c) VHDL module defining waveguide entities/architectures and *optoelectronic simulation package*. The optical power generated by a laser diode is linearly proportional to the forward bias current of laser driver when the bias current is larger than the threshold current. Therefore the forward bias current from the laser diode SPICE model can be linearly related to the optical power in the VHDL models. Similarly photo currents for the photodiodes can also be linearly formulated by using the known responsivity of the optical receiver. Optical output signals

```vhdl
-- Waveguide_Bending.vhdl file to define bending waveguide entities

USE WORK.Optical_package.ALL;
USE STD.TEXTIO.ALL;

ENTITY Waveguide_Bending IS
        GENERIC( wg_width, bending_radius : length;
                 prop_loss : db_per_length;
                 excess_loss : db_loss;
                 dispersion_m, dispersion_w : dispersion;
                 index_core, index_buffer, index_effective : REAL;
                 noiseIndex : REAL
        );
        PORT( Op1 : IN Optical; Op2 : OUT Optical);
END Waveguide_Bending;

ARCHITECTURE behavior OF Waveguide_Bending IS

    BEGIN

      PROCESS (Op1)
        VARIABLE lambda, w, R : REAL;
        VARIABLE wg_length : length;
        FILE testOUT : TEXT IS out ''bending.out'';
        VARIABLE outLine : LINE;

        BEGIN
          Op2.spectral_width <= Op1.spectral_width;
          Op2.wavelength <= Op1.wavelength;
          IF length'POS(Op1.wavelength) = 0 THEN
            lambda := REAL(length'POS(1.5 um))/1000.0;
          ELSE
            lambda := REAL(length'POS(Op1.wavelength))/1000.0;
          END IF;
          w := REAL(length'POS(wg_width))/1000.0;
          R := REAL(length'POS(bending_radius))/1000.0;
          wg_length := 2.0 * PI * bending_radius / 4.0;
          Op2.intensity
              <= TRANSPORT bending(Op1.intensity, wg_length,
                                   prop_loss, excess_loss,
                                   index_effective, index_buffer,
                                   index_core,
                                   lambda, w, R
                                  )
                             + noiseIndex * RAND
                    AFTER delay(wg_length, index_core);

          IF Op1.intensity = 0.0 THEN
            Op2.rise_time <= 0 ns;
          ELSE
            Op2.rise_time
                <= bandwidth( Op1.spectral_width,
                              wg_length,
                              dispersion_m, dispersion_w
                            )
                    + Op1.rise_time;
          END IF;
      END PROCESS;

END behavior;

CONFIGURATION wg_bending  OF Waveguide_Bending IS
        FOR behavior
        END FOR;
END wg_bending;
```

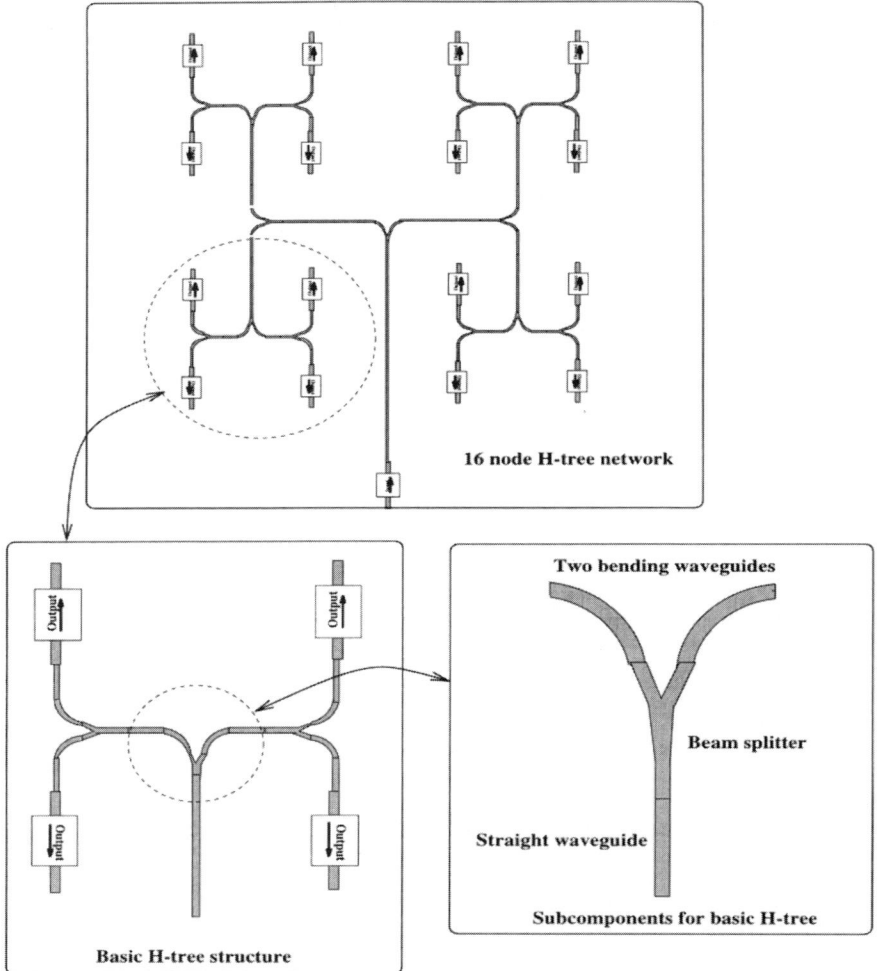

Fig. 3. 16 node H-tree networks using basic VHDL waveguide entities.

Fig. 4. VHDL simulation modules for optical waveguide.

Table 1. Description of physical OCDN parameters on MCM for VHDL simulations.

1 Optical signal specifications

signal wavelength	1.3 um
spectral width	1 nm

2 MCM specifications

distribution network structure	symmetric H-tree
Number of nodes	sixteen
MCM size	$2.4 \times 2.4\,\text{cm}^2$
MCM substrate	silicon

3 Optical waveguide specifications

waveguide type	single mode, step index waveguide
excess loss	0.1 dB/component
waveguide propagation loss	0.1 dB/cm
waveguide bending radius	2 mm
material dispersion	r ps/nm km
waveguide dispersion	2 ps/nm km
waveguide material	silica glass

4 I/O coupling methods

laser-to-fiber coupling	butt coupling using GRIN lens
fiber-to-waveguide coupling	butt coupling
waveguide-to-photodetector	micromachined silicon mirror

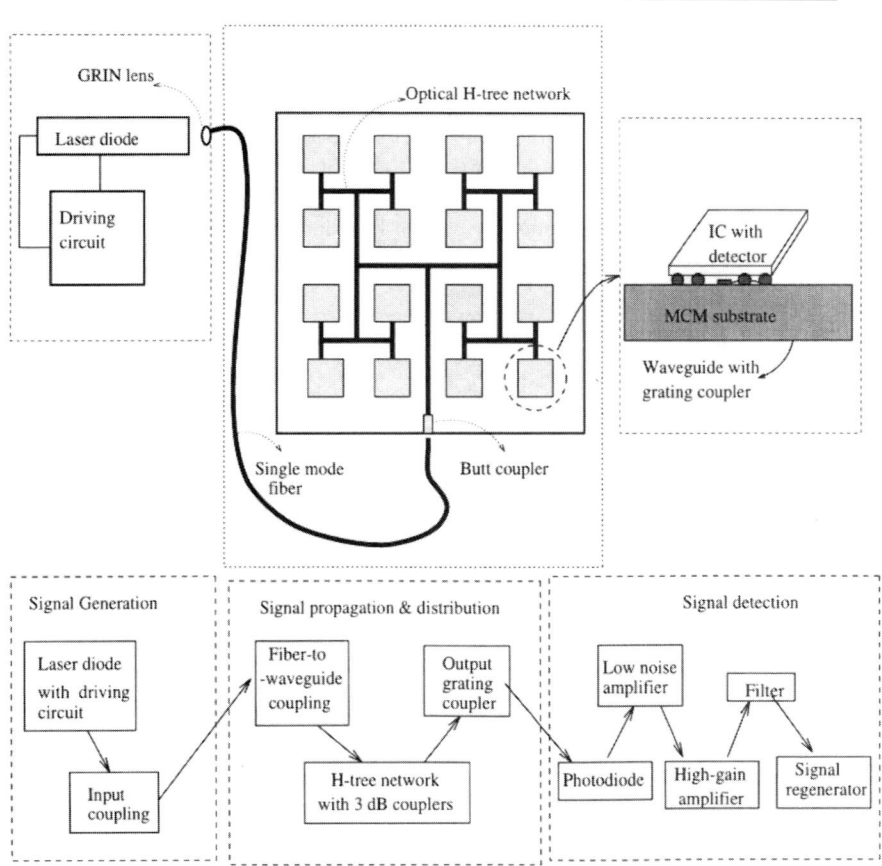

Fig. 5. Functional block diagram of optoelectronic clock distribution network (OCDN) on multichip module (MCM).

from a SPICE module for a laser diode transmitter is fed into a VHDL simulation module and these signals are propagated and distributed along the network to generate output signals at 16 fanout nodes. These signals are fed back into a SPICE module of a photodiode receiver to generate an electrical signal. Fig. 5 illustrates a functional block diagram of OCDN on MCMs. In this example, VHDL models incorporates the SPICE simulation results for their own simulations by translating the transient current response to the VHDL optical input signal profiles. Using the same idea, the VHDL optical output profiles can also be translated to SPICE input waveform for their subsequent electrical circuit simulations.

During the VHDL simulation C module calculates optical signal behaviors numerically by providing values of bending loss, signal propagation loss, and rise time degradation due to dispersion. Fig. 6 shows an example VHDL simulation result of 16 node OCDN on MCM. This illustrates an input/output signal levels of an optical beam using the information in Table 1. This example also assumes zero background noise.

4. Conclusion

Optoelectronic interconnect requires an integration of optoelectronic and electronic devices at chip, module, or board levels. The integration of optoelectronic devices into the conventional electronic information processing systems has to be justified in term of cost, performance, and portability of existing system software and hardware. A VHDL-base system level optoelectronic interconnect simulation methodology is described in this paper. This methodology provides a possibility of proliferation of optoelectronic interconnection technologies into the existing computer or communication systems. Successful modeling and simulation is performed for 16 node OCDN on MCMs and this provides vital system design information useful for power budget, system bandwidth, or power consumption analysis. Additional VHDL models like X junction, wavelength division (de)multiplexer (WD(D)M) and demultiplexer are currently under development to include every possible waveguide structures. Technology information for different waveguide materials and waveguide structures are also under investigation to build a waveguide technology library for VHDL simulation purposes. Once a complete optoelectronic device design library is built, rapid prototyping of optoelectronic interconnect system would be possible by synthesizing complex optical waveguide structures as EDA tools synthesize complex electronic systems. Simulation of multi-wavelength optical fiber communication links can be effectively supported by modifying the *Optical* signal definition to represent multiple signals of different wavelengths and providing proper VHDL entity/architecture combinations of optical fibers, WDM and WDDM.

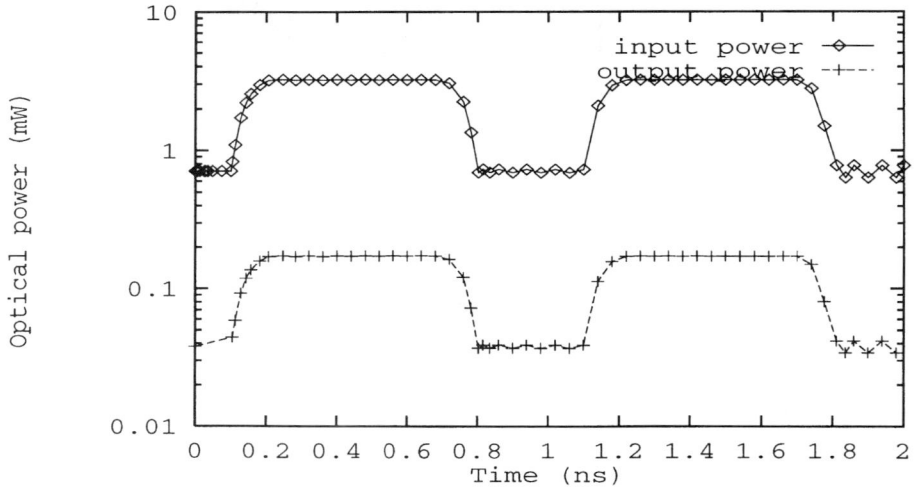

Fig. 6. VHDL simulation results of 16 node optical H-tree clock distribution network on MCM showing Input/Output optical power relationship.

References

1. R. Goyal. "Managing signal integrity." In *IEEE Spectrum*, pp. 54–58, March, 1994.
2. J. L. Hennessy and D. A. Patterson. *Computer Architecture: A Quantitative Approach*, Morgan Kaufmann Publishers, Inc. 1990.
3. J. W. Goodman, F. I. Leonberger, S. Y. Kung, and R. A. Athale. "Optical interconnections for VLSI systems." *Proceedings of the IEEE* 72(7), pp. 850–866, 1984.
4. S. Koh, H. W. Cater, and J. T. Boyd. "Synchronous global clock distribution on multichip modules using optical waveguides." *Optical Engineering* 33(5), pp. 1587–1595, 1994.
5. D. L. Lee. "Electromagnetic Principles of Integrated Optics." New York: Wiley, 1986.
6. R. G. Hunsperger. "Integrated Optics: Theory and Technology." Springer-Verlag, Third Edition, 1991.
7. H. Nishihara, M. Haruna, and T. Suhara. "Optical Integrated Circuits." McGraw-Hill, 1989.
8. Seungug Koh, Harold W. Carter, and Joseph T. Boyd. "Performance comparison of global clock distribution networks on multichip modules based on electrical and optical interconnect technologies." *Proceedings of the SPIE Optoelectronics Conference*, 1995.
9. A. Fisher and H. T. Kung. "Synchronizing Large VLSI Processor Arrays." *IEEE Transaction on Computers* c-34(8), 1985.

Seungug Koh received the B.S. in Physics from Korea University, Seoul, Korea in 1988 and the M.S. in Applied Optics from Rose-Hulman Institute of Technology, Terre Haute, Indiana, in 1990, and the Ph.D. in Computer Engineering from University of Cincinnati, Cincinnati, Ohio in 1994. He is currently an assistant professor of Electrical and Computer Engineering at the University of Dayton, Dayton, Ohio. His research interest includes Physical Design of VLSI and MCMs Systems, Distributed and Parallel Computer Architecture, VHDL Modeling, Simulation and Synthesis of Photonic Integrated Circuits and MEMS, Optical Link Simulation using VHDL for All-optical Networks, and Integrated Optoelectronics and MicroOptoElectroMechanical Systems (MOEMS). In 1994 he received the "Best Ph.D. Dissertation Award" from ECE department of University of Cincinnati and he also obtained a U.S. patent in 1995.

Hierarchical Analog Behavioral Modeling of Artificial Neural Networks

MONA M. AHMED, HISHAM HADDARA AND HANI F. RAGAIE
ECE Department, Faculty of Engineering, Ain Shams University, Cairo

Received February 28, 1996; Accepted March 27, 1997

Abstract. A hierarchical methodology for analog behavioral modeling of the basic building blocks of neural networks is presented using HDL-A.[1] This hierarchy is formed of three levels in order to satisfy the different requirements of the CAD tools which may incorporate the models. The presented models include all the nonidealities present in the actual circuit in addition to being flexible and consuming shorter simulation time. This improvement in simulation time is verified through examples at both the circuit and system levels.

Key Words: behavioral modeling, neural networks, analog VLSI

1. Introduction

A behavioral model is a mathematical, algorithmic, or pictorial representation of a thing or concept in terms of its characterizing parameters [1]. The concept of behavioral modeling has been efficiently utilized in the digital circuits domain leading to a considerable reduction in the design cycle [2]. When being applied to the analog domain, several difficulties were encountered mainly due to the fact that high level analog functions are simple and highly technology and circuit dependent which contradicts with the abstractness and technology independence of the modeling concept. Nevertheless behavioral models incorporated in libraries for systems, utilizing the analog VLSI technology as a major technology such as artificial neural networks (ANNs), can enhance the activities in the field of analog design automation which still lags behind its digital counterpart.

Analog VLSI is one of three technologies used for hardware implementation of ANNs. Its suitability arises from the small computational circuits and the high speed which it is capable of providing, thus leading to fast and area efficient ANN systems with low power dissipation [3]. In addition to the fact that some of the traditional analog design requirements such as accurate absolute component values, device matching, precise time constant, etc., are not major concerns. This is primarily because computation precision of individual neurons does not seem to be of paramount importance. Design factors such as signal handling and frequency range are not well defined while the problem of interconnection, compactness, and power dissipation represent design hurdles [4].

Having all above features in addition to a modular structure formed of a limited set of components, the analog basic blocks utilized in neural networks applications can be combined together in a single library that can be embedded in any analog CAD environment. Each block is characterized by being simple, reconfigurable, and versatile, thus any proposed model should be as simple as the circuit and consuming a shorter simulation time.

In this paper, we propose a hierarchical methodology for the modeling of analog primitive cells and apply it to the case of ANNs. The objective of this work is to develop a library for the analog cells used in neural networks. In the next section the structure of artificial neural networks is illustrated, while Section 3 discusses the applicability of the modeling concept to the neural networks case. Section 4 presents the proposed modeling approach and Section 5 emphasizes it through examples at the circuit and the system levels together with the simulation results. Finally conclusions are drawn in Section 6.

2. Artificial Neural Networks Basic Building Blocks

The basic structure of a neural network consists of a matrix of synapse cells interconnecting an array of input neurons with an array of output neurons as shown in Fig. 1a. The inputs are multiplied by weight values in the synapses, and the results of multiplication are summed and compared with the threshold values in the output neurons. In addition to this modular architecture, ANNs are characterized by their learning capability, whereby synaptic strengths between neurons are adaptively changed according to a specific algorithm. Consequently, the basic building blocks of dedicated analog hardware realization of neural networks are divided into circuits for multiplication and adaptation for synaptic operation, circuits for summation and thresholding (nonlinearities) for neuron's operation (refer to Fig. 1b), in addition to other specific circuits depending on the type of the network as the Gaussian synapse, and the Winner Take All (WTA) circuit [5–15].

3. Behavioral Modeling of Neural Networks

Digital CAD tools use behavioral modeling at design, simulation, and validation phases. Analog circuits

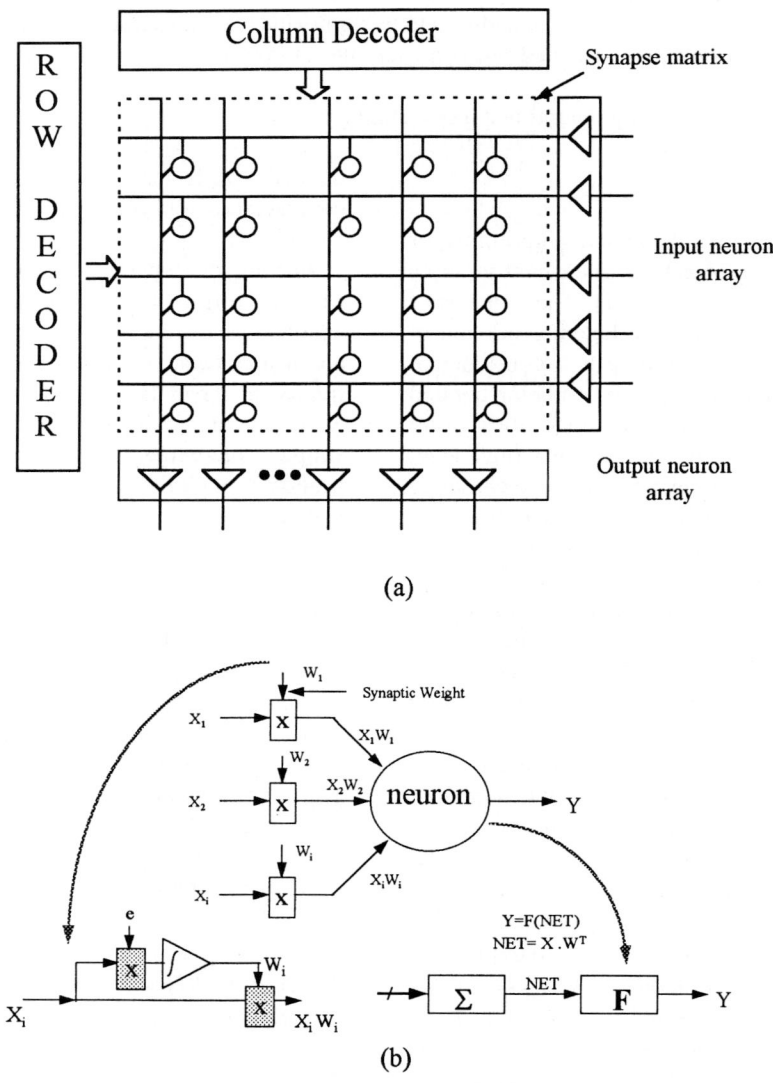

Fig. 1. (a) General purpose neural network configuration. (b) Basic artificial neuron model illustrating the synapse weight adaptation and the internal structure of the cell body.

have not yet mature CAD tools as those of their digital counterpart [16].

The field of ANNs is one of those fields employing analog VLSI technology in addition to its nature being formed of a limited set of modules. Applying the concept of behavioral modeling to neural networks may have some restrictions.

- ANNs are still in a development phase lacking well approved circuit architectures.
- The model should be independent of the nature of the CAD tools utilizing it.
- The model should be closer to the circuit level in order to be distinguished from the algorithms used for software simulation of ANNs.
- Analog libraries in general are difficult to implement due to the fact that the primitive analog cells have more than one function with a number of nonidealities depending on the used technology.

The above points can be counteracted by:

- The trend in neural networks is towards modular and reconfigurable architectures which are well suited for the modeling concept.
- The number of CAD tools depending on the behavioral description of any system as its entry are those used at the behavioral simulation phase, the validation phase, and possible synthesis phase, whose requirements can be gathered in a single model.
- The level of accuracy of the model and the degree of closeness to the circuit level is mainly user defined.
- In order to capture all the circuit functions, the model could be formed of a single entity, describing the interface of the circuit and a number of architectures, describing its function.

4. A Proposed Modeling Approach

In order to satisfy the four points previously mentioned, we propose a hierarchical methodology for the modeling of ANNs. The same point has been the subject of previous publications [17–19]. In this work, the model, written in analog hardware description language (HDL-A) supported by the electrical simulator ELDO[2], is composed of three hierarchical levels. This decomposition is important for providing flexibility in the choice of the desired degree of accuracy depending on the application. The first level is a block diagram representation of the circuit serving as a first order functionality check, in fact this level is approximately the same as the mathematical algorithm of the ANNs except for having an interface related to the physical circuit. The second level incorporates all the nonidealities of the circuit. This level is the most important one due to its adherence to the circuit technique utilized together with the fact that it is parameterized thus providing both fast and accurate simulation.

Improvement in the design cycle can be achieved using the second level, as the designer can first simulate the system using the model in order to optimize the different circuit parameters, and finally realize these parameters using the actual circuit [17].

The last level involves statistical modeling as it includes statistical information about the circuit in addition to errors arising from the variation in its characterizing parameters resulting from technological process. These errors could be found by applying Monte Carlo analysis while varying the technological parameters and visualizing its effect on the circuit [18]. The last level should be provided for each used technology. Contradiction with the technology independence of the modeling concept is encountered in the last level but in the analog case, this effect is extremely important especially if the model will be further processed by any synthesis tool that selects the circuit satisfying the system requirements with the smallest acceptable error.

5. Results and Discussion

In this section, we will focus on the first and the second hierarchical levels of the model, while the last level will be treated in future phases of work. Three primitive cells, the wide range Gilbert multiplier [20], the Gaussian circuit utilized in the radial basis function neural network (RBFNN) [21], and the Winner Take All (WTA) circuit [4,21] are presented and their circuits and models are compared regarding simulation time and accuracy. A macromodel composed of the basic primitives (Transconductance element, I-V converter, and Hard-Limiter) and describing the operation of the 4 bit Hopfield A/D converter is given to emphasize the power of the modeling concept at the system level.

5.1. Wide Range Gilbert Multiplier

The Gilbert multiplier circuit produces an output current proportional to the product of the two differential input signals $(V_1 - V_2)$ and $(V_3 - V_4)$. Thus the model in its first level represents the functionality of this circuit simply as a multiplication process independent of the operating regime or the circuit specific parameters but preserving the nature of the input–output signals as shown in equation 1:

$$I_{out} = G_m(V_1 - V_2)^*(V_3 - V_4) \qquad (1)$$

The Gilbert multiplier circuit has been implemented in different ways and technologies [15,20,21]. The one published by Mead [15] (Fig. 2a) has the advantage of operating in the subthreshold regime thus consuming low power. The output current of this circuit is given as

$$I_{out} = I_b \tanh\left(\frac{n(V_1 - V_2)}{2KT/q}\right)$$
$$\times \tanh\left(\frac{n(V_3 - V_4)}{2KT/q}\right) \qquad (2)$$

Where I_b is the bias current provided by the transistor Mb, n is the reciprocal of the subthreshold slope. The model in its second level deals with a specific circuit with specific nonidealities, and thus it encounters two problems. The first is that the equations used in the model ACHITECTURE should be a function of the external characterizing parameters which the user senses independent of any technological parameter, therefore instead of including n, and I_b, we include the linear operational range represented by the parameter V_{sat} and the transconductance gain G_m given in the following equations:

$$G_m = \frac{I_b}{(2KT/qn)^2}, V_{sat} \approx \frac{2KT}{qn} \qquad (3)$$

The second problem lies in the contradiction between having a parameter G_m as an external parameter, set by the user, and a biasing PIN, whose voltage affects the value of G_m. The solution is to consider the parameterized property of the model as a priority over having an identical interface as that of the actual circuit.

The main nonidealities present in the circuit are the output conductance G_{out} depending the channel length modulation parameter of the output transistors, the common mode rejection ratio (CMRR) and the offset voltage(V_{off}) depending on any mismatch present in the input transistors of the differential pairs due to fabrication processes. These nonidealities are introduces in equation 2 as:

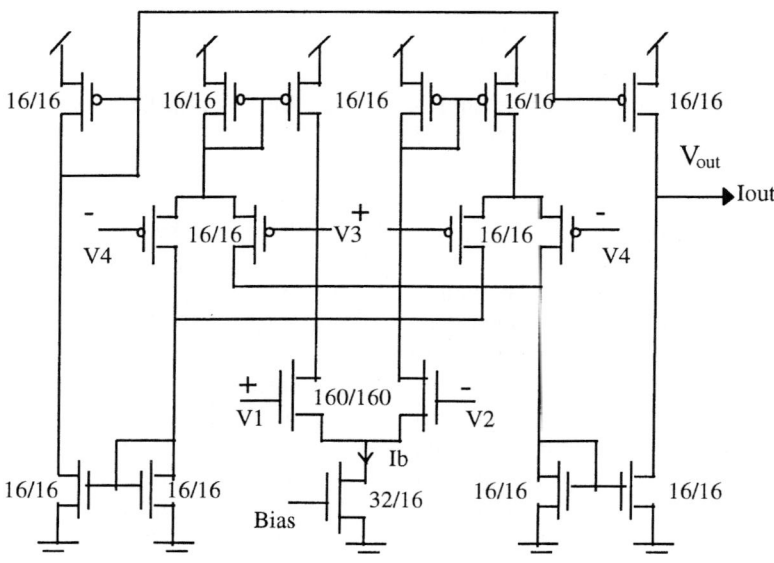

Fig. 2. Circuit Schematic of the wide range Gilbert multiplier.

$$I_{out} = G_m \tanh\left(\frac{V_1 - V_2 + V_{off1}}{V_{sat}}\right)$$
$$\times \tanh\left(\frac{V_3 - V_4 + V_{off2}}{V_{sat}}\right)$$
$$+ \text{CMRR1} \tanh\left(\frac{V_1 + V_2}{2V_{sat}}\right)$$
$$+ \text{CMRR2} \tanh\left(\frac{V_3 + V_4}{2V_{sat}}\right)$$
$$- G_{out} V_{out} \qquad (4)$$

The Gilbert multiplier circuit is known to have a restriction on the values of the output voltage which may be imposed on it when being employed in any system. For small values of output voltages below (V_{min}), the output current suffers a large increase while the opposite occurs for large values of voltage approaching that of the supply (V_{max}) [15]. This is mainly due to the inversion in the roles of source and drain of the output MOS transistors (refer to Appendix I[A]).

The simulation results (Fig. 3a,b) for the model

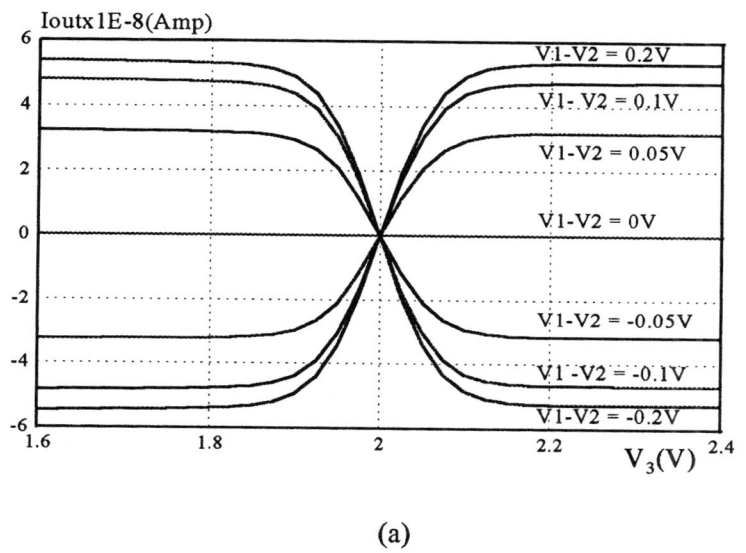

(a)

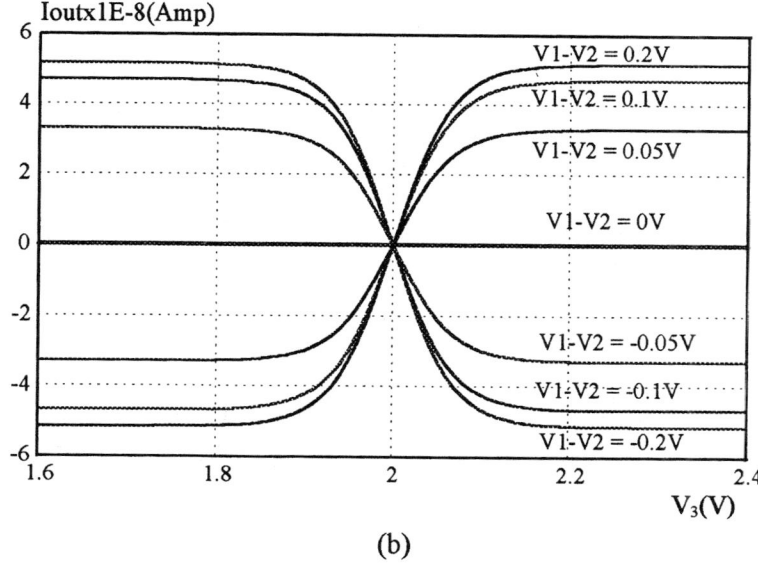

(b)

Fig. 3. Simulation results for the output current of the Gilbert multiplier. (a) Circuit. (b) Model.

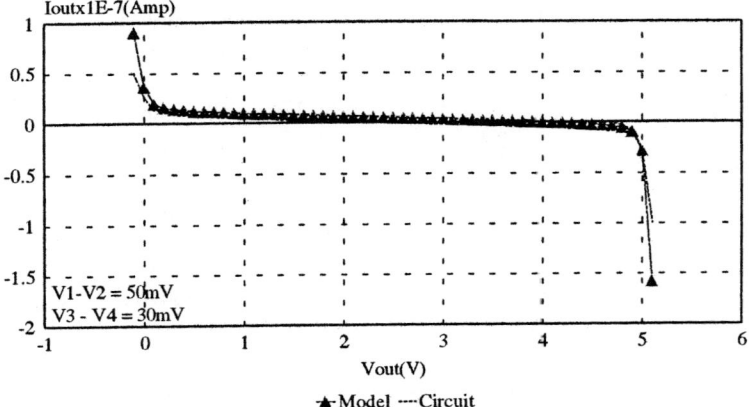

Fig. 4. Comparison between the circuit and model of the wide range Gilbert multiplier output characteristics.

and the circuit represent the output current of the multiplier as a function of the second differential voltage ($V_3 - V_4$) for different values of the first differential voltage ($V_1 - V_2$). The results of the model agree to that of the circuit with 2.6% average error.

Nonideal effects as the finite output conductance and the variation of currents at small and large output voltages are shown in Fig. 4. The equations describing the current for output voltages below (V_{min}) and above (V_{max}) are empirical because high accuracy is not required when operating outside the useful range of operation.

If we define the improvement in simulation time as the ratio between the difference in simulation time of the circuit and the model to the simulation time of the circuit under the same simulation conditions regarding the type of analysis and the number of points, then 43% improvement is achieved when performing DC analysis.

5.2. The Gaussian Function Circuit

The radial basis function neural network (RBFNN) and the probabilistic neural network (PNN), used in classification and functional approximation [3], employ synapses that do not just perform the multiplication process. In the PNN, the synapses are Gaussian function circuits such that the summation of their output currents produces a Gaussian mixture that can approximate the probability density function (PDF) of any class of data [18]. The circuit shown in Fig. 5a produces an output current approximately satisfying the Gaussian function when fed by a differential input voltage ($V_{in} - V_W$). The Gaussian shape of the transfer characteristics is mainly due to this differential nature. The output current is given as:

$$I_{out} = AI_x - (I_3 + I_4) \qquad (5)$$

Where A is the drain current ratio in the current mirror. When $V_{in} = V_w$, the summation of the two currents ($I_3 + I_4$) is zero, and the total output current equals AI_x. As the difference between the two inputs increases within the linear operational region of the circuit, either I_3 or I_4 increases at the expense of the other, decreasing the overall current. This takes place until the summation of the two currents saturates and the output current is kept at a constant value. Thus the standard deviation of the approximate Gaussian function mainly depends on the linear range of the differential amplifier.

As previously stated, the model in its first level is simply the basic function of the circuit given in equation 6

$$I_{out} = I_{peak} e^{-(V_{in}-V_w)^2/2\sigma^2} \qquad (6)$$

Where I_{peak} is the peak value of the Gaussian shaped output current, V_{in} is the input voltage, V_w is the voltage describing the weight of the synapse, and σ is the standard deviation of the obtained characteristics.

In order to model the specific circuit shown in Fig. 5a, we have to include all the deviations from the ideal characteristics and thus the effect of current saturation has to be considered. In addition to the two parameters I_{peak} and σ, a new parameter V_{th} is defined as the voltage after which the current saturates. This effect is

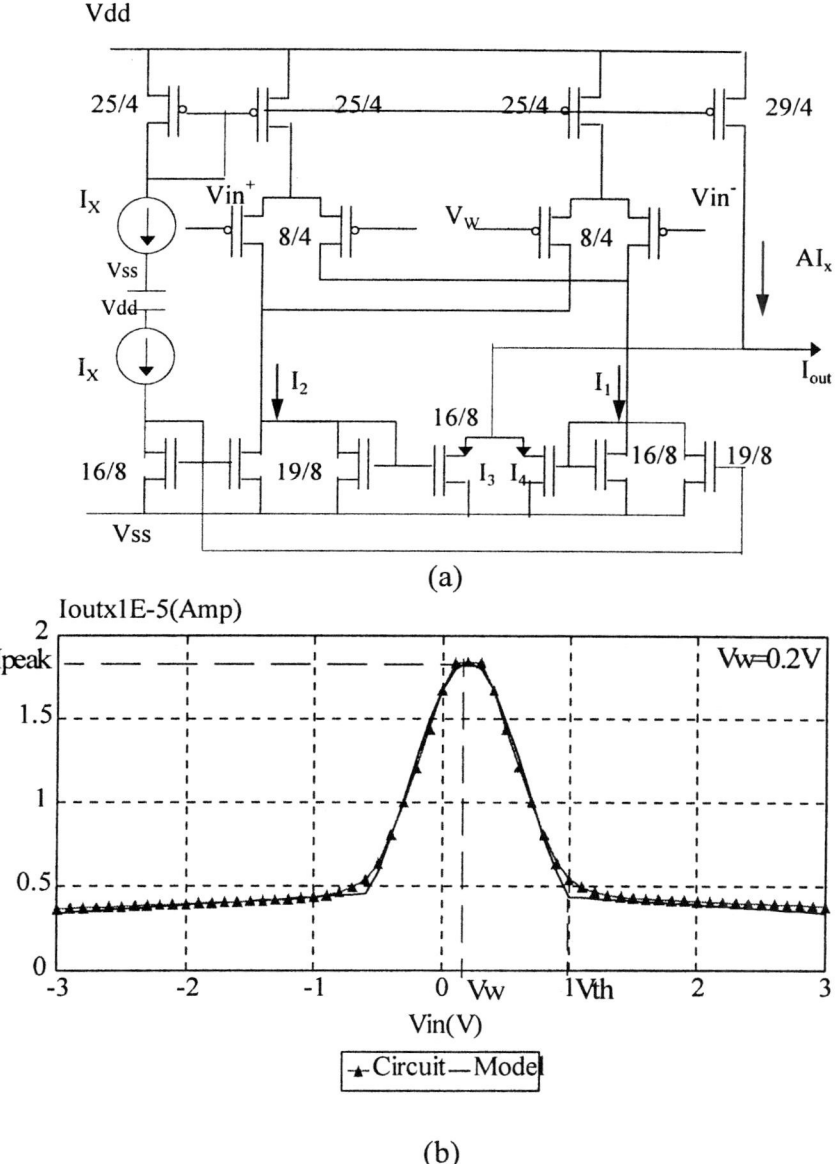

Fig. 5. (a) Circuit schematic of the Gaussian function synapse. (b) Comparison between the output current resulting from the simulation using the circuit and the model of the Gaussian function synapse.

modeled by multiplying equation 5 with an arbitrary function $m(v)$. This function is constant inside the operating Gaussian range and thus the output current is a pure Gaussian function of the input voltage. Outside this range, the funtion $m(v)$ tends to make the current changes its slope to acquire a linear behavior (refer to Appendix I[B]). Simulation results of the circuit and the model are shown in Fig. 5b. The model approximately satisfies the function of the circuit with 4.3% average error and 72% improvement in simulation time on performing DC analysis.

The second realization, shown in Fig. 6, in which the whole network performs the Gaussian function is used in RBFNN [22]. The equation governing the operation of the network is given as

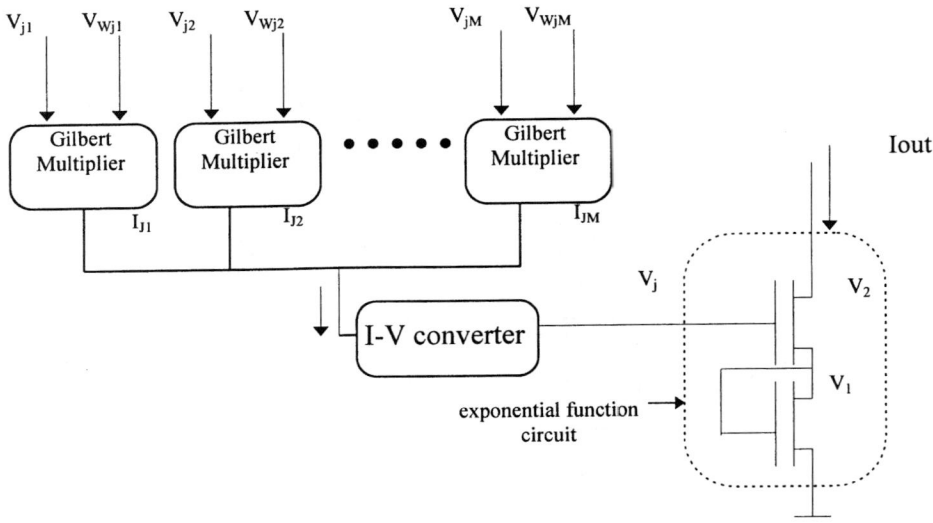

Fig. 6. Schematic diagram of the Gaussian function network.

$$I_{out} = I_0 e^{-nK/n+1 \sum G_m (V_{in}-V_w)^2} \quad (7)$$

Where the squaring of the differential input $V_{in} - V_w$ is performed using a wide range Gilbert multiplier with a transconductance gain G_m. The output current from all multipliers are summed and converted into voltage using an I-V converter satisfying equation 8

$$V_{out} = V_{ref} - KI_{in} \quad (8)$$

Where V_{ref} is a reference voltage, I_{in} is the input current, and K is the conversion gain. The output voltage from the I-V converter is then fed to an exponential function circuit formed of two transistors operating in the subthreshold regime, where n is the reciprocal of the subthreshold slope.

The model in this case may be treated as a description of the functionality of the whole network, and thus it will have the same ideal Gaussian function equation of the previous model, or we can consider it as a macromodel formed of three primitives (Gilbert multiplier, I-V converter, and an exponential circuit). Modeling in this case is completely related to the circuit architecture and thus it loses its abstractness. The main parameters of the three models are the transconductance of the Gilbert multiplier G_m, the conversion gain K and the constant reference voltage V_{ref} of the I-V converter, a constant current I_o and a factor $A = n/n + 1$, related to the parameter σ in the first model, for the exponential function.

The macromodel shows larger simulation time as compared to that of the first model. Its simulation results are given in Fig. 7 showing the output from each stage and satisfying 33.8% improvement in simulation time when performing DC analysis with 7.1% average error due to the accumulation of errors from each stage.

5.3. *The Winner Take All (WTA) Circuit*

The conventional view of the artificial neuron as a number of weighted interconnection, and a processing nonlinear element is violated in special neural networks architectures that employ different circuits in a manner different from the usual perspective. The winner take all circuit is one of these. For competitive neural network algorithms, such as Kohonen's self organizing feature map, Hamming network, or learning vector quantization (LVQ) networks, it is required to find the most active unit out of a neuron layer.

The function of the WTA cell is to accept input signals, compare their values, and then produce a high output corresponding to the largest signal while all other outputs are set to a low value.

The circuit shown in Fig. 8a is a typical implementation that has been reported operating at strong [21], and weak inversion regions [4]. In [23] the same circuit was used but it operates in the current mode not the transresistance mode. The idea in all

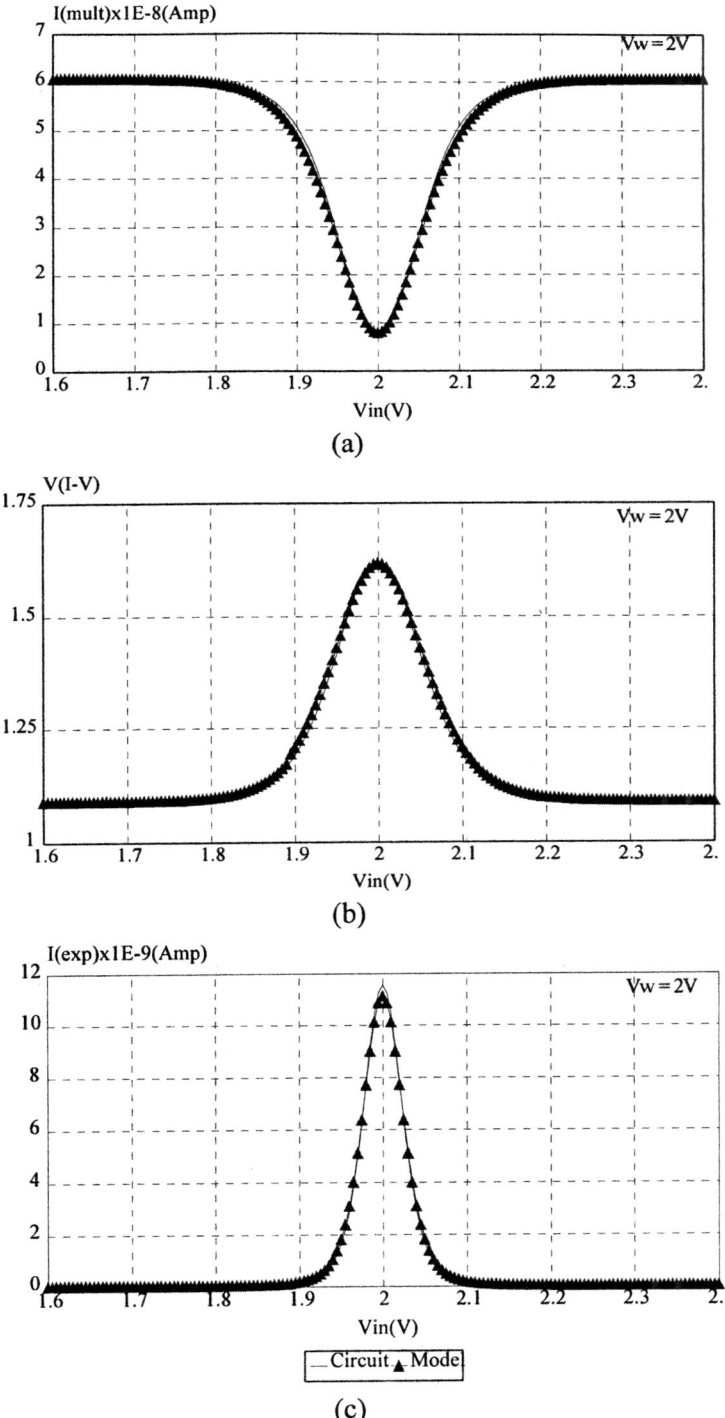

Fig. 7. Comparison between the simulation results for the hierarchical circuit and model describing the Gaussian function network performance. (a) The output current of the Gilbert multiplier. (b) The output voltage of the I-V converter. (c) The output current of the exponential function.

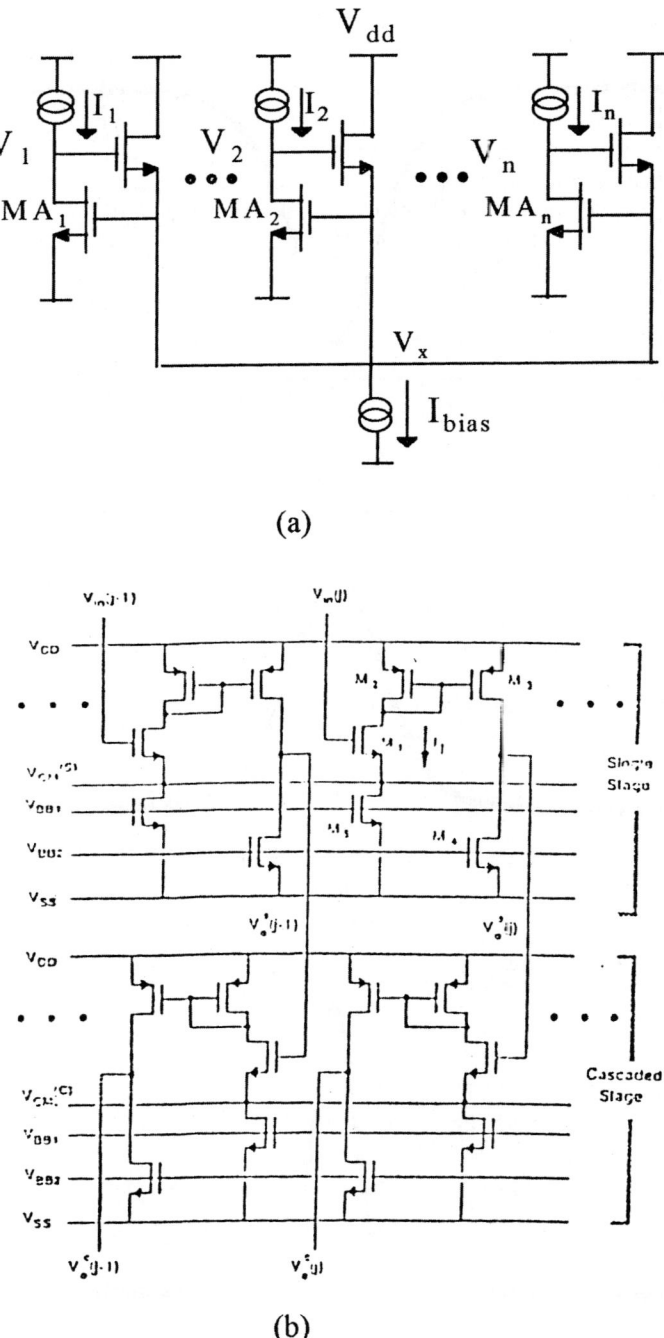

Fig. 8. Circuit schematic of the Winner Take ALL (WTA) cell. (a) Current mode operation. (b) Voltage mode operation.

cases is the same, having N inputs, for the largest current I_K the drain voltage of transistor MA_k should be larger than the other transistors in order to accommodate the largest input current since all transistors have identical gate-source voltages of $V_x - V_{ss}$. Thus the largest input current I_k can produce the largest node voltage V_k as the output voltage. Notice that the inputs and outputs actually share a common node. The circuit is tested by using 2 branches only, and applying two currents. The current I_2 is fixed at 6 μA, and the current I_1 varies from 5.5 μA to 6.5 μA. The results for DC analysis is shown in Fig. 9a. The dimensions of all transistors is 2 μm/20 μm and the bias current is 10 μA.

The second implementation, shown in Fig. 8b, is completely operating at the voltage mode but using the same concept of having all source terminals of the input transistors tied together. Since the source terminals are at the same potential for all the cells, the current flowing through each cell is related to the square of the input voltage. Thus, the strongest input can secure the largest amount of current out of the total bias current. This largest current is converted and amplified to produce the largest voltage as the output of the winning node. If the input voltage differences are sufficiently large, the winner output is saturated at the positive power supply value, while the other outputs are saturated at the negative power supply value [21]. The DC analysis simulation results are shown in Fig. 9b.

Modeling of the WTA circuit can not be done by simply stating the circuit equations, and defining the nonidealities. The number of inputs and outputs in this case is a variable, since we have N inputs. The model should be completely abstract to just capture the circuit functionality.

The parameters of the model are:
1. The number of inputs and outputs (N).
2. The response time or the delay during the transition from high to low level or vice-versa (res_time).
3. The high output value (V_{high}).
4. The low output value (V_{low}).

The main difference between the models of the two

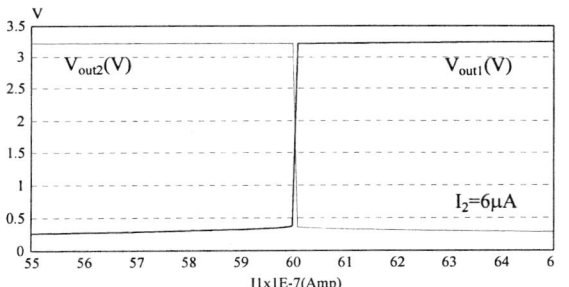

(a)

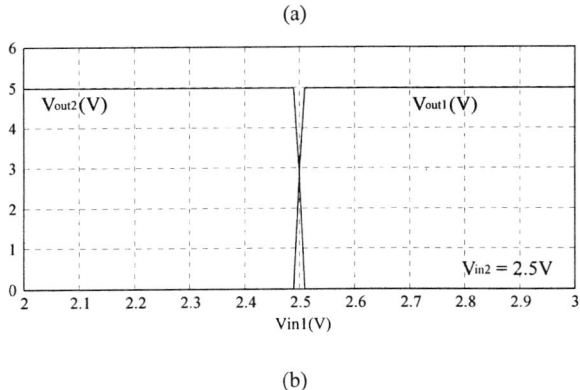

(b)

Fig. 9. DC simulation results for a 2 branches WTA cell.
(a) Current mode circuit. (b) Voltage mode circuit.

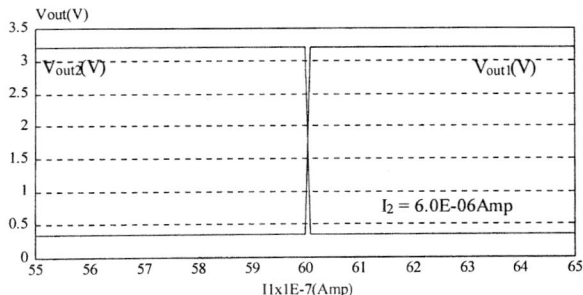

(a)

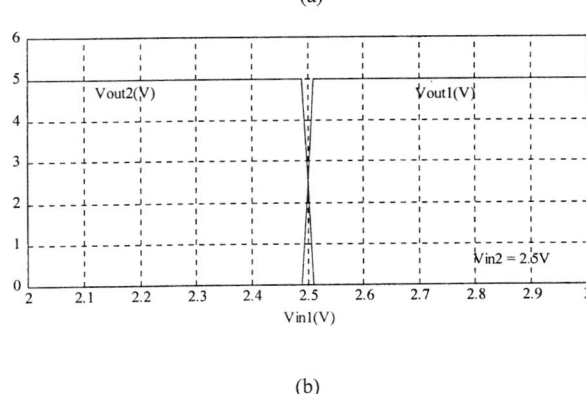

(b)

Fig. 10. DC simulation results for a 2 branches WTA cell.
(a) Current mode model. (b) Voltage mode model.

circuits previously mentioned is that in the first circuit, we don't have inputs and outputs, rather a single group of nodes that act as both inputs and outputs. The high and low values for the second implementation are VDD, and VSS, while for the first, it depends on the dimensions of the transistors and thus it is a parameter that should be given by the user. The critical problem in the model for the first implementation is the nature of the input, that is current. Circuit and behavioral model simulators, such as SPICE, and ELDO, solve the circuit or the model by solving KVL and consequently producing the current in all branches, the reverse situation usually produces convergence problems that the model will never be executed. The only solution available at this time is to read the current in an *EQUATION* block [24] and assign the values of the currents to internal parameters. This solution increases the model simulation time. The model of the second circuit is relatively more efficient, the circuit after cascading has a large number of transistors that the model simulation time is small compared to that of the circuit specially that working in the voltage mode does not require any additional steps or equations. The DC analysis for the model of the first and the second circuits is shown on Fig. 10. It is worth noting that the efficiency of the model will be much higher in both cases if the number of branches increases as the model will not be changed while the circuit size will increase thus increasing its simulation time. Operating at the current mode produces an average error equals 2.87% with an improvement in simulation time of 18.6%, while that at the voltage mode produces an average error that does not exceed 0.22% with an improvement in simulation time of 64.8%.

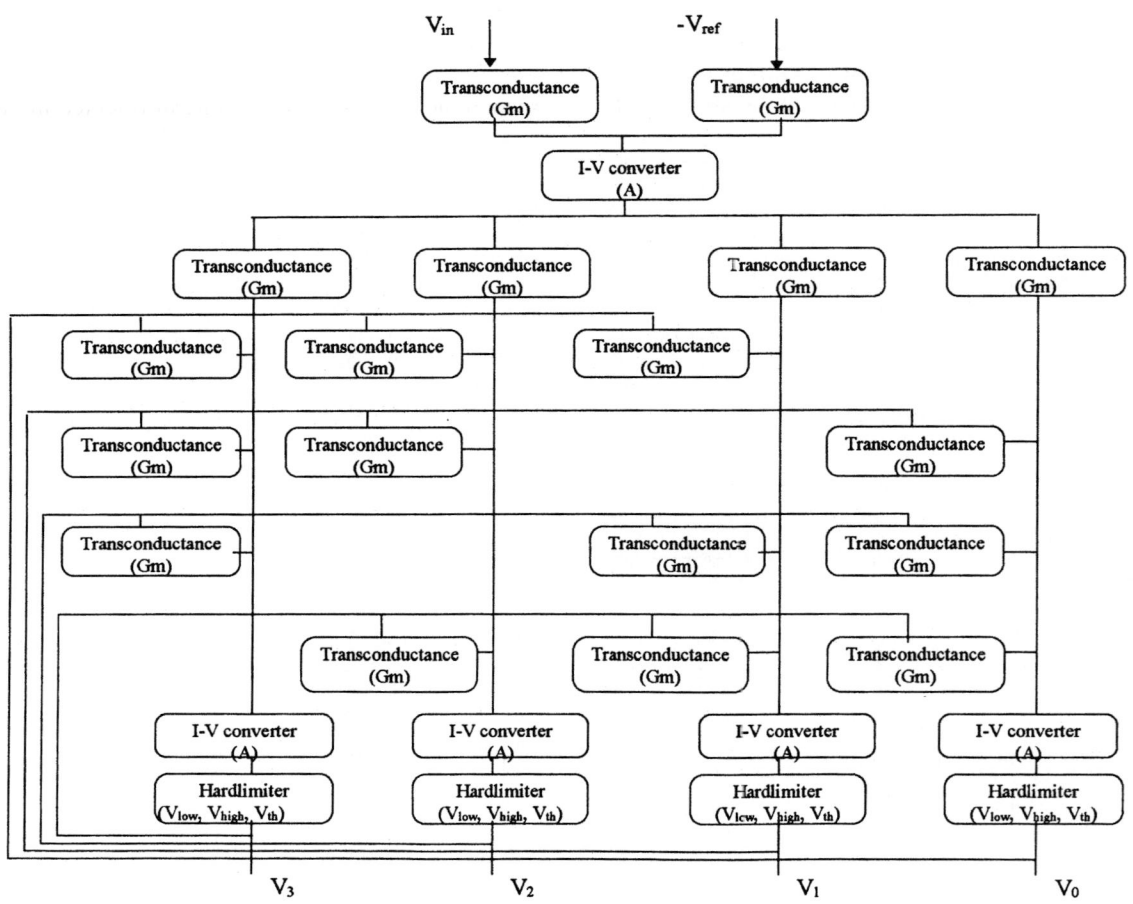

Fig. 11. Block diagram for the 4 bit Hopfield A/D converter [25].

5.4. The Hopfield A/D Converter

This circuit has gained its popularity in the mid-1980s thanks to its simplicity and regular structure. It has become now of limited importance due to the problems associated with its implementation and learning [4]. It is given as an example in this section just for the illustration of the model capabilities for high connectivity at the subsystem level. The 4 bit Hopfield A/D converter can be implemented in different ways [25,26]. The methodology used in [26] is well suited for the modeling process as the system is constructed of three major blocks (transconductance element, I-V converter, and hard-limiter). A block diagram for the circuit is shown in Fig. 11 with the three primitives described by their characterizing parameters. The parameters G_m and K have been described in the previous subsection, V_{low} and V_{high} are the two levels of the hard-limiter, and V_{th} is the threshold voltage. The number of transistors used for each block is small, thus the improvement in simulation time will not be noticeable (20% under DC analysis). The main improvement in this system when utilizing the models is the fact that the designer can change the parameters without going into the details of the circuit. Results provided by the model and the circuit are given in Fig. 12.

6. Conclusion

In this paper a methodology is given for the analog behavioral modeling of ANN. Among the overall applications utilizing the analog VLSI technology, ANN seems to be the most suitable to be modeled and afterwards synthesized. The models given at the second hierarchical level are circuit oriented in order to capture the closeness to the hardware implementation. We can not compare the simulation time improvement for different circuit sizes because each model has its own degree of complexity independent of the size of the circuit. Simulation results of the given examples have shown noticeable degree of improvement in simulation time which varies from one circuit to another with acceptable accuracy.

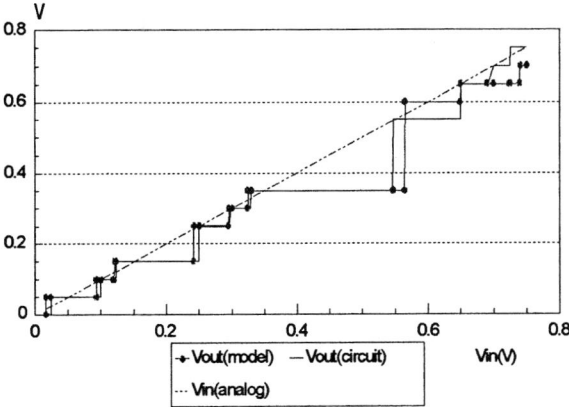

Fig. 12. Comparison between the simulation results of the circuit and the model of the 4 bit Hofield A/D converter.

Appendix I

A. The Architecture of the HDL-A Model of the Wide Range Gilbert Multiplier

```
ARCHITECTURE functional1 of Gilbert_multW is
VARIABLE vout_min : analog ; —The min. output voltage after which the current
                              —increases as function of the output voltage.
VARIABLE vout_max : analog ; —The max. output voltage after which the current
                              —decreases as function of the output voltage.
VARIABLE K : analog ;         —A technology parameter indicating the
                              —subthreshold slope.
BEGIN

RELATION
```

Appendix A. (Continued)
PROCEDURAL FOR INIT = >

```
Gm := 5.6E-7;        —the transconductance is function of the bias current.
Gout := 6.6E-10;     —the output conductance
voff1 := 25.0E-3;    —the offset voltage of the first differential pair
voff2 := 49.0E-3;    —the offset voltage of the second differential pair
CMRR1 := 200.0;      —The common mode rejection ratio of the first differential pair
CMRR2 := 140.0;      —The common mode rejection ratio of the second differential
                       pair.
vsat := 43.0E-03;    —A Technology parameter related to the subthreshold slope
PROCEDURAL FOR DC, TRANSIENT, AC = >
k := 2.0*vsat;
IF ((in1_pos.v < 0.0) and (in1_neg.v < 0.0)) OR ((in2_pos.v < 0.0) and
(in2_neg.v < 0.0)) then
REPORT ''OUT OF RANGE'' SEVERITY ERROR;
ELSE
            IF in2_pos.v = > in2_neg.v then
                vout_min := 0.33*((in2_neg.v) - bias.v);
                else
                vout_min := 0.33*((in2_pos.v) - bias.v);
                end if;
                vout_max := 0.99*(vdd.v-0.1);
                if (vout_min < outp.v) and (outp.v < vout_max) then
                outp.i% = -(K*Gm*th((in1_pos.v-in1_neg.v+voff1)/
k)*th((in2_pos.v-in2_neg.vvoff2)/k)+Gm/(10.0**(CMRR1/20.0))*th((in1_pos.v
+in1_neg.v)/k)+Gm/(10.0**(CMRR2/20.0))*th((in2_pos.v+in2_neg.v)/k)-
Gout*outp.v);
        else if (outp.v > = vout_max) then

    if   (K*Gm*th((in1_pos.v-in1_neg.v+voff1)/k)*th((in2_pos.v-
in2_neg.vvoff2)/k)+Gm/(10.0**(CMRR1/20.0))*th((in1_pos.v+in1_neg.v)/
k)+Gm/(10.0**(CMRR2/20.0))*th((in2_pos.v+in2_neg.v)/k) - Gout*outp.v) > 0.0
then
       outp.i% = -(K*Gm*th((in1_pos.v-in1_neg.v + voff1)/k)*th((in2_pos.v-
in2_neg.v - voff2)/k)+Gm/(10.0**(CMRR1/20.0))*th((in1_pos.v+in1_neg.v)/
k)+Gm/(10.0**(CMRR2/20.0))*th((in2_pos.v+in2_neg.v)/k)-Gout*outp.v)*(2.0
- exp(((-vout_max +outp.v)/0.02*K)*((-vout_max +outp.v)/0.02*K)));

           else
    outp.i% = -(K*Gm*th((in1_pos.v-in1_neg.v + voff1)/k)*th((in2_p((in2_
pos.v-in2_neg.v - voff2)/k) +Gm/(10.0**(CMRR1/20.0))*th((in1_pos.v+in1_
neg.v)/k)+Gm/(10.0**(CMRR2/20.0))*th((in2_pos.v+in2_neg.v)/k)-Gout*outp.
v)*(exp(((-vout_max +outp.v)/0.02*K)*((-vout_max +outp.v)/0.02*K)));

            end if;

        else
        if (K*Gm*th((in1_pos.v-in1_neg.v + voff1)/k)*th((in2_pos.v-
```

Hierarchical Analog 135

Appendix A. (Continued)
```
in2_neg.v - voff2)/k) +Gm/(10.0**(CMRR1/20.0))*th(in1_pos.v+in1_neg.v) +Gm/
(10.0**(CMRR2/20.0))*th(in2_pos.v+in2_neg.v) - Gout*outp.v) > 0.0 then
        outp.i %= -(k*Gm*th((in1_pos.v-in1_neg.v + voff1)/
vsat)*th((in2_pos.v-in2_neg.v-voff2)/k) +Gm/(10.0**(CMRR1/
20.0))*th((in1_pos.v+in1_neg.v)/k) +Gm/(10.0**(CMRR2/
20.0))*th((in2_pos.v+in2_neg.v)/k) - Gout*outp.v)*(exp((vout_min -outp.v)/
0.01*k));

            else
    outp.i %= -(vsat*Gm*th((in1_pos.v-in1_neg.v + voff1)/vsat) -
Gout*outp.v)*th((in2_pos.v-in2_neg.vvoff2)/k) +Gm/(10.0**(CMRR1/
20.0))*th((in1_pos.v+in1_neg.v)/k) +Gm/(10.0**(CMRR2/
20.0))*th((in2_pos.v+in2_neg.v)/k)*(2.0 - exp((vout_min -outp.v)/0.01*k));
            end if;
            end if;
end if;
end if;
end relation;
END ARCHITECTURE functional1;
```

B. The Architecture of the HDL-A Model of the Gaussian Function Synapse

```
ARCHITECTURE functional1 of guassian is
variable m : analog;
—A function to be multiplied by the ideal current to account for nonidealities
variable vth2 : analog;
BEGIN
RELATION
PROCEDURAL FOR INIT =>
Ipeak := 10.0E-9;
sigma := 10.0E-1;
Voff := 0.0;
vth := 1.5;
PROCEDURAL FOR DC, TRANSIENT =>
vth2 := 2.0*in_weight.v - vth;
        IF (in_signal.v > vth2) and (in_signal.v < vth) then
            m := 1.0;
        ELSE
            m := exp(-k*((vth - in_weight.v)/sigma)*((vth - in_weight.v)/
sigma))* exp(k*((in_signal.v -    in_weight.v)/sigma)*((in_signal.v -
in_weight.v)/sigma))*(1.0 -0.1*abs(in_signal.v));
        END IF;
OUTP.I %= -m*Ipeak* exp(-k*((in_signal.v - in_weight.v)/sigma)*((in_signal.v -
in_weight.v)/sigma));
—The outut current from the pin(outp) is given by its ideal equation multiplied by
the function m(v)
End Architecture functional1
```

C. The Architecture of the HDL-A Model of WTA Cell

```
ARCHITECTURE functional1 OF wta IS

variable ll : analog;
variable ss1 : integer;
variable tt1 : integer;
state tt : analog;
state ss : analog;
variable vv : integer;
variable vx : integer;
state temp : analog;
state temp1 : analog;
state temp2 : analog;
   BEGIN

 RELATION

PROCEDURAL FOR INIT =>
response_time := 1.5e-05;
vv := 1;
tt := 1.0;
ss := 2.0;
temp1 := 0.0;

PROCEDURAL FOR DC, AC =>
for x in 1 to p-1 loop
if inp(x).v = inp(x+1).v then
outp(x).v %= vdd.v/2.0;
outp(x+1).v %= vdd.v/2.0;
else
        for x in 1 to P loop
        if inp(vv).v < inp(x).v then
        vv := x;
        else
        vv := vv;
        end if;
        outp(x).v %= vss.v;
        end loop;
outp(vv).v %= vdd.v;
end if;
end loop;

PROCEDURAL FOR TRANSIENT =>
        for x in 1 to P loop
        if inp(vv).v < inp(x).v then
        vv := x;
        else
        vv := vv;
        end if;
```

Appendix C. (Continued)
```
        temp := real(vv);
outp(x).v %= vss.v;

        end loop;
tt := real(vv);

outp(vv).v %= vdd.v;

if tt /= previous(tt) then
ss := previous(tt);
ss1 := integer(ss);
tt1 := integer(tt);
end if;
if temp1 < previous(temp1) then

outp(ss1).v %= temp1;
outp(tt1).v %= 5.0 - temp1;

else
if temp1 > previous(temp1) then
outp(ss1).v %= 5.0 - temp1;

outp(tt1).v %= temp1;
else
if temp1 = previous(temp1) then
outp(ss1).v %= vss.v;
outp(tt1).v %= vdd.v;
end if;
end if;
end if;

        for x in 2 to P loop
if inp(x).v > inp(x-1).v then
ll := vss.v;
else
ll := vdd.v;
end if;
end loop;
equation(temp1) for transient =>
temp1 + ddt(temp1)*response_time == ll;
END RELATION;
END ARCHITECTURE functional1;
```

Notes

1. HDL-A is a VHDL based analog language developed by ANACAD Electrical Engineering Software.
2. ELDO is a trademark of ANACAD Electrical Engineering Software.

References

1. H. Mantooth and M. Fiegenbaum. *Modeling with an Analog Hardware Description Language*. Kluwer, Norwell, MA, 1995.
2. E. Liu and A. Vincentelli. "Behavioral representations for VCO and Detectors in Phase-Lock Systems." *IEEE 1992 Custom Integrated Circuits Conference*.
3. M. Glesner and W. Pöchmüller. *Neurocomputers An Overview of Neural Networks in VLSI*. Chapman & Hall, London, 1994.
4. C. Mead and M. Ismail, eds. *Analog VLSI Implementation of Neural Systems*. Kluwer, Norwell, MA, 1990.
5. Jacek M. Zurada. *Introduction to Artificial Neural systems*. West Publishing company, 1992.
6. A. Cichocki and R. Unbehauen. *Neural Networks for Optimization and Signal Processing*. Wiley, New York, 1993.
7. J. Van der Spiegel, P. Mueller, D. Blackman, P. Chance, C. Donham, R. Etienne-Cummings, and P. Kinget. "An analog neural computer with modular architecture for real-time dynamic computations." *IEEE J. Solid-State Circuits* SC-27, pp. 82–92, 1992.
8. S. Gowda, B. Sheu, J. Choi, C. Hwang, and J. Cable. "Design and Characterization of Analog VLSI Neural Network Modules." *IEEE J. Solid-State Circuits* 28, pp. 301–312, 1993.
9. J. Choi, S. Bang, and B. Sheu. "A programmable Analog VLSI Neural Network Processor for Communication Receivers." *IEEE Trans. Neural Networks* 4, pp. 484–494, 1993.
10. A. Andreou, K. Boahen, P. Pouliquen, A. Pavasovic, R. Jenkins, and K. Strohbehn. "Current-Mode Subthreshold MOS Circuits for Analog VLSI Neural Systems." *IEEE Trans. Neural Networks* 2, pp. 205–213, 1991.
11. B. Barranco, E. Sanchez-Sinencio, A. Rodriguez-Vazquez, and J. Huertas. "A Modular T-Mode Design Approach for Analog Neural Network Hardware Implementations." *IEEE J. Solid-State Circuits* 27, pp. 701–712, 1992.
12. F. Kub, K. Moon, I. Mack, and F. Long. "Programmable Analog Vector-Matrix Multipliers." *IEEE J. Solid-State Circuits* 25, pp. 207–214, 1990.
13. P. Hollis and J. Paulos. "Artificial Neural Networks Using MOS Analog Multipliers." *IEEE J. Solid-State Circuits* 25, pp. 849–855, 1990.
14. N. Saxena and J. Clark. "A Four-Quadrant CMOS Analog Multiplier for Analog Neural Networks." *IEEE J. Solid-State Circuits* 29, pp. 746–749, 1994.
15. C. Mead. *Analog VLSI and Neural Systems*. Addison-Wesley, Reading, MA, 1989.
16. M. Ismail and J. Franca, eds. *Introduction to Analog VLSI Design Automation*. Kluwer, Norwell, MA, 1990.
17. A. M.Abdelatty, H. Haddara, and H. F. Ragaie. "Analog Behavioral Modeling of Artificial Neural Networks." *Proc. ICM'94* pp. 140–143.
18. M. Elmasry, eds. *VLSI Artificial Neural Networks Engineering*. Kluwer, Norwell, MA, 1994.
19. E. Sanchez-Sinencio. "A Unified Enviroment for High-Performance Analog Signal Processing." *Proc. ICM'94*, pp. S49–S49.
20. Randall L. Geiger, Phillip E. Allen, and Noel R. Strader. *VLSI Design Techniques for Analog and Digital Circuits*. McGraw Hill, 1990.
21. B. Sheu and J. Choi. *Neural Information Processing and VLSI*. Kluwer, Norwell, MA, 1995.
22. S. Watkins, P. Chau, and R. Tawel. "A radial basis function neurocomputer implemented with analog VLSI circuits." *Proc. IEEE/INNS Inter. Joint Conf. Neural Networks* 2, pp. 607–612, 1992.
23. A. Andreou, K. Boahen, P. Pouliquen. A. Pavasovic, and K. Strohbehn. "Current-Mode Subthrehold MOS Circuits for Analog VLSI Neural Systems." *IEEE Trans. Neural networks* 2, pp. 205–213, 1991.
24. HDL-A user manual.
25. B. Kosko. *Neural Networks for Signal Processing*. Prentice Hall, 1992.
26. M. Tan. "Synthesis of Artificial Neural Networks by Transconductances Only." *Analog Integrated Circuits and Signal Processing* 1, pp. 339–350, 1991.

Mona Mostafa Ahmed received the B.Sc. and M.Sc. degrees from Ain Shams University, Cairo, Egypt in 1993 and 1996 respectively. From 1993 to 1996 she was a teaching and research assistant in the same university. Currently she is a graduate fellow in the Ohio-State university, USA where she is working towards her Ph.D. in analog VLSI design.

Hani Ragaie received the B.Sc. and M.Sc. degrees from Ain Shams University, Cairo, Egypt in 1967 and 1972 respectively. He received the "Doctorat de Specialite" and the "Doctoral d'Etat" in electronics from Grenoble University, France in 1976 and 1980 respectively. From 1972 to 1979 he was a research assistant at LETI and then at THOMSON-EFCIS from 1979–1980, both in Grenoble, France. He then joined the Electronics and Communication Engineering Department, Faculty of Engineering, Ain Shams University as assistant professor. Since 1990, he is a professor of electronics in the same department where he is also the director of "Integrated Circuits Lab". His research interests include modeling and characterization of MOS devices, mixed-mode VLSI design, electrochemistry of silicon surfaces and porous-silicon devices.

Fault Modeling and Simulation Using VHDL-AMS

A. J. PERKINS, M. ZWOLINSKI, C. D. CHALK AND B. R. WILKINS
Department of Electronics and Computer Science, University of Southampton, Southampton SO17 1BJ, UK

Received February 28, 1996; Accepted June 19, 1997

Abstract. Fault simulation is an accepted part of the test generation procedure for digital circuits. With complex analog and mixed-signal integrated circuits, such techniques must now be extended. Analog simulation is slow and fault simulation can be prohibitively expensive because of the large number of potential faults. We describe how the number of faults to be simulated in an analog circuit can be reduced by fault collapsing, and how the simulation time can be reduced by behavioral modeling of fault-free and faulty circuit blocks. These behavioral models can be implemented in SPICE or in VHDL-AMS and we discuss the merits of each approach. VHDL-AMS does potentially offer advantages in tackling this problem, but there are a number of computational difficulties to be overcome.

Key Words: analog simulation, analog test, fault simulation, fault modeling, analog VHDL

1. Introduction

As integrated circuits have grown in complexity, the importance of testing has also grown. It is not sufficient to consign test issues to some post-design phase. Design-for-test, in order to facilitate testing at the manufacturing stage, is now firmly established as an important aspect of digital system design [1]. For analog and mixed-signal circuits, however, testing is still often regarded as a peripheral matter, because design-for-test is perceived to adversely affect performance. Moreover, the testing of small analog circuits may simply be a number of functional tests. Large analog and mixed-signal integrated circuits make functional testing very difficult. Individual parts of the design cannot be tested in isolation, while functional tests may not reveal deeply embedded defects.

While functional testing can be and is used for digital circuits, it is common to assume the existence of stuck-at, bridging and open faults. The testing methodology therefore becomes one of identifying the presence or otherwise of these *structural* faults [2,3]. The object of a testing program for a circuit is to verify whether or not a fault exists using the smallest possible number of test vectors. In general, one test will detect more than one fault and each fault is covered by more than one test. Test pattern generation is thus the process of selecting an optimal set of tests from all possible input patterns. As the generation of the optimal set is unlikely to be feasible for anything other than the smallest circuits, algorithms such as the D-algorithm or PODEM are used to find a test pattern for one fault [2]. Alternatively, random test patterns may be applied. Once a pattern is found, its fault cover can be assessed and all faults detectable with that pattern can be dropped from further consideration. The assessment of fault cover is made using *fault simulation*. For each fault, a copy of the circuit is made containing that fault and no other. The original, fault-free circuit and the faulty copies are simulated with the given test pattern. If the output of the faulty copy differs from that of the original the fault is detectable using that test pattern. Techniques, such as *concurrent fault simulation* [2], exploit the fact that the differences in behavior between the faulty and fault-free circuits are often relatively small, and by avoiding redundant element evaluation, reduce the computational effort required to evaluate all the faulty circuit copies.

Most of these techniques are not immediately applicable to analog circuits. Structural defects, such as open and short circuits, can be identified, but these do not necessarily manifest themselves a simple stuck faults. Hence, test vector generation has to be done in an *ad hoc* manner. The major difficulty, however, is fault simulation. The simulation of analog circuits is at least two orders of magnitude slower than that of similarly sized digital circuits [4]. The number of potential faults to be modeled is far greater in an analog circuit. In a digital circuit, we are only concerned with the interconnection between gates and we assume that the nodes will be stuck-at 1 or stuck-at 0. In an analog circuit we must concern ourselves with every node in the circuit and, in the worst case—in the absence of detailed layout information—we must assume the possibility of every pair of node-to-node shorts and of each branch being open circuit. Thus the number of fault simulations to be performed is likely to be greater than for a similar digital circuit. Finally, a digital node is, after transitions have stabilized and contentions have been resolved, either 1 or 0; an analog node has a value represented by a floating point number. A node in a faulty digital circuit is either exactly the same as in the fault-free circuit or different, making redundancy easy to exploit. In an analog circuit a slight difference between a faulty and a fault-free node may result in a massive difference elsewhere in the circuit, thus making redundancy very difficult to exploit. Moreover, the existence of faults in the analog circuit model may take component models outside their normal, characterized region of behavior and ultimately may render the faulty circuit unsimulatable.

It can thus be seen that analog fault simulation has not been a usable tool for analyzing the effectiveness of test vectors, because too many slow simulations are required. The motivation for the work described in this paper has been to test the effectiveness of various supply current monitoring techniques in detecting analog faults and hence the need to perform fast analog fault simulations.

We first describe how the speed of an analog simulation may be increased by modeling circuits behaviorally. Macromodels for standard analog building blocks such as operational amplifiers have been developed using controlled sources in SPICE. VHDL-AMS offers an alternative approach that is in many ways simpler [5]. We further describe how simpler behavioral models can be used for operational amplifiers in closed-loop configurations. This is important for fault modeling because open-loop characteristics are not generally observable in embedded circuits. The number of simulations can be reduced by observing which faults cause similar observable behavior. An example of fault collapsing is given in Section 3. By modifying the characteristics of the fault-free SPICE and VHDL-AMS behavioral models, faulty behavior can be modeled. We therefore go on to show that such behavioral models can emulate the transistor level models very accurately. One particular difficulty addressed is the propagation of faulty behavior through fault-free circuit blocks; this can take those fault-free blocks outside their characterized region of behavior, thus requiring more complex models.

In order to assess the suitabilty of VHDL-AMS for this work, a commercial implementation of a draft of the 1076.1 working documents was used. Certain difficulties with simulation speed and with robustness were encountered, and these are discussed.

2. Behavioral Modeling

A simple two-stage CMOS operational amplifier is shown in Fig. 1. This is significantly simpler than a "real" design; nevertheless it has 8 transistors, 9 nodes and 15 branches, and in comparison to a digital circuit of a similar size, the time required for simulation is relatively high. A number of SPICE macromodels for general operational amplifiers have been developed [6]; the model described in [7] includes the supply current. A comparison of simulation times is given in Table 1. The macromodels are all based around controlled sources. In SPICE, controlled sources may be simply linear functions of one voltage or current, or polynomial functions of several voltages or currents. Such elements are therefore not sufficient to model voltage or current limiting as happens for instance when an op-amp's output voltage saturates. Such effects are instead modeled by the inclusion of, for instance, diodes. Similarly, differential inputs may be modeled by a simplified MOS transistor pair. SPICE macromodels of circuits such as op-amps are therefore a combination of controlled sources and simplified semiconductor elements.

Unsurprisingly, the characterization of such macromodels is not easy. In effect, we are forced to do a multidimensional optimization. In practice, the problem is simplified, to some extent, because each part of the macromodel corresponds to some observable behavior, so the optimization can be done in parts. Analog hardware description languages make such macromodeling a potentially simpler task.

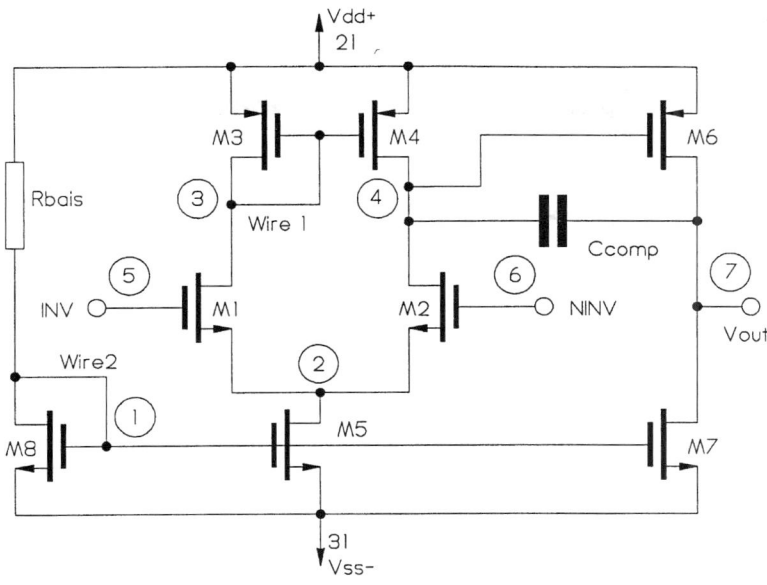

Fig. 1. Two stage CMOS op-amp.

Instead of forcing a mathematical model onto a limited set of defined components, the model can be implemented directly. A number of VHDL-AMS models (or to be precise, models based on the implementation of the draft) have been written which illustrate this, Fig. 2 shows a SPICE and VHDL-AMS implementation of a voltage limiter. For the SPICE model a number of standard components are required to achieve the desired function where as in the VHDL-AMS implementation a simpler mathematical description is used. Another example of a VHDL-AMS model, this time of an open loop op-amp, is shown in Fig. 3, and in Fig. 4 the complete VHDL-AMS code is given. The behavior of the op-amp is defined in terms of its gain, transfer function, output impedance, slew rate, CMRR and power supply current characteristics. Each of these characteristics can be incorporated directly into a mathematical expression. The derivation of the model is described in the Appendix. As with the macromodel of [7], the op-amp model includes the supply current variation. Unlike the SPICE macromodel, the behavior of the model maps directly onto the block diagram.

Again, unlike SPICE, the relation between each parameter and the model is explicitly stated, without the need to map the effects onto controlled source parameters. Because of this clarity, the VHDL-AMS model is, arguably, superior to the SPICE model. Op-amps are not usually used in open-loop configuration, therefore it may not be necessary to model open-loop effects that are not observable in closed-loop configurations. This is especially significant with respect to testing—a common testability metric is whether a node that might have a fault can be controlled from an input and whether the effect of that fault can be observed at an output. Thus it may be argued that a change in the open-loop gain, for

Table 1. Comparison of simulation times for various CMOS op-amps.

Op-amp model	CPU time (s) for 30 μs simulated time
Transistor level, two-stage	15.6
Transistor level, cascode	23.7
Complete macromodel	9.7
Macromodel without V & I limiting	6.6

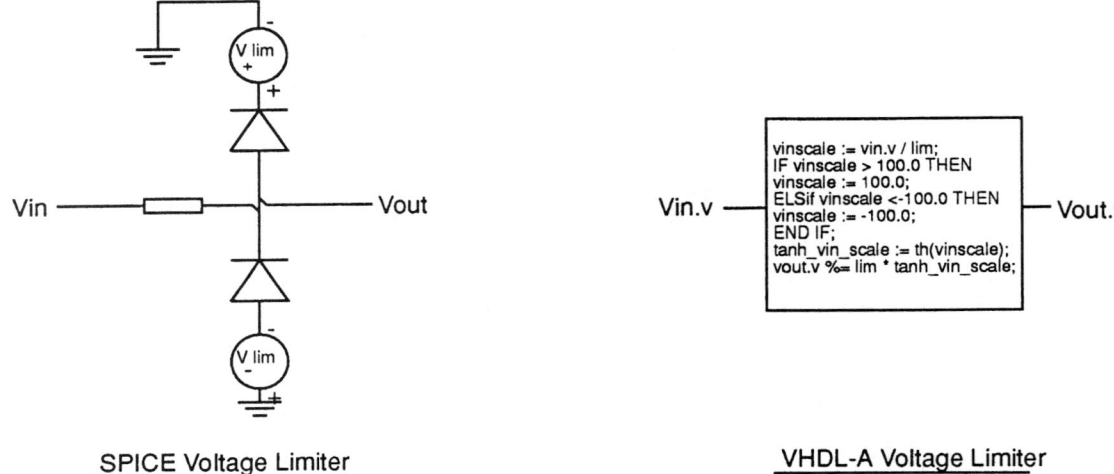

Fig. 2. SPICE and VHDL-AMS limiters.

instance, may not be directly observable in a closed-loop configuration, and that therefore there is no point in attempting to model it. If an op-amp is modeled in its closed-loop configuration, its macromodel can be significantly simplified. Fig. 5 shows the macromodel for an op-amp configured as an inverting amplifier. It will be observed that 7 parameters are sufficient to completely model the behavior of this circuit. The same macromodel can be used for a non-inverting amplifier configuration, but with different parameter values. A similar model can be constructed for the summing amplifier case. A VHDL-AMS version of the model is shown in Fig. 6. As can be seen, this model is significantly simpler than the open-loop model of Fig. 3. Fig. 7 compares two simulations of the audio mixer circuit of Fig. 8 for both the output voltage and the supply current. In one simulation the circuit is modeled entirely at the transistor level and in the other using VHDL-AMS models of each block. The differences between the models are insignificant.

3. Fault Collapsing and Fault Modeling

As has been noted, defects in digital circuits can often be characterized as single stuck-faults. This fault model represents a form of fault collapsing as a

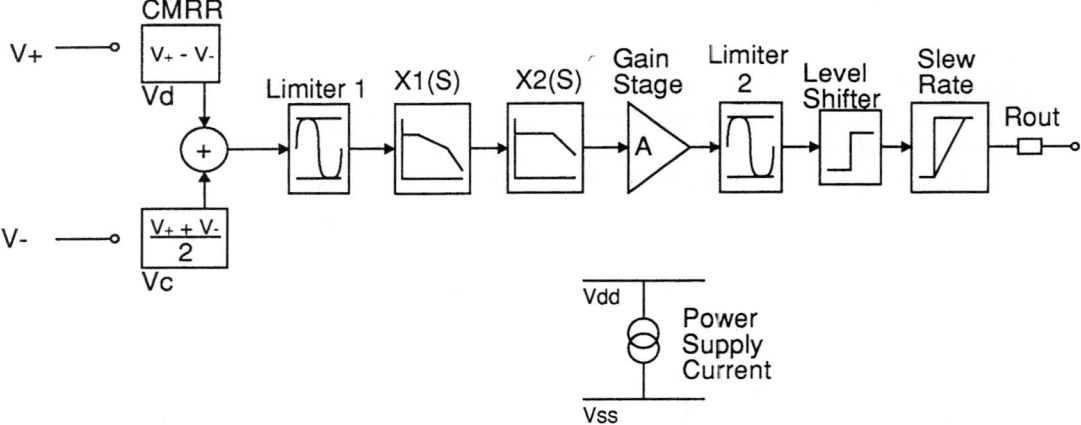

Fig. 3. VHDL-A op-amp block diagram.

Fault Modeling and Simulation Using VHDL-AMS   145

```
-- op-18.vhdla Full opamp version 2
ENTITY opamp18 IS
    GENERIC (p1,p2,z1,gain,cmrr,srate,rop,ron,rdc,gmdd,limdiff : real);
    PIN (vinp,vinn,vout,Vdd,Vss : electrical);
END ENTITY opamp18;

ARCHITECTURE a OF opamp18 IS
    STATE   outp1,outp2,soutp1,soutp2,sinp1,
            sinp2,inp1,inp2 : analog;           -- Implicit Variables
    VARIABLE ip1,ip2,iz1 :analog;               -- temp poles and zero
    VARIABLE vd,vc,vca,vin,vzeroa,comgain :analog;  -- CMRR
    VARIABLE inl1,outl1,inl2,outl2,lim1,lim2    -- Limiters in/out
           ,vinscale,tanh_vin_scale : ANALOG;   -- Limiter internal
    VARIABLE outg :analog;
    VARIABLE inls,outls : analog;               -- level shift
    VARIABLE insr,rate,diff,diff2,msrate :ANALOG; -- Slew rate
    STATE    outsr :Analog;                      -- slew rate
    VARIABLE inro,io,ro :ANALOG;
    VARIABLE idc,iac :ANALOG;
BEGIN
  RELATION
    PROCEDURAL FOR init =>
    -- default generics
    p1 := 1.0e7;                    -- low frequency pole
    p2 := 1.0e2;                    -- high frequency pole
    z1 := 1.0e8;                    -- high frequency zero
    gain := 1.0e6;                  -- open loop DC gain
    cmrr := 90.0;                   -- input CMRR
    srate := 500.0e3;               -- slew rate
    rop := 1.0;                     -- output resistance when vout +
    ron := 1.0;                     -- output resistance when vout -
    rdc := 5.0e3;                   -- Power supply offset current resistance
    gmdd := 100.0e-6;               -- Power supply current vout conductance
    limdiff := 0.5;                 -- Output voltage limit from Vdd

    -- seting up varables
    ip1 := 1.0/(twopi*p1);
    ip2 := 1.0/(twopi*p2);
    iz1 := 1.0/(twopi*z1);
    comgain := 1.0 / (10.0** (cmrr/20.0) );

--************************************************************;
-- ********* AC & DC *********************************;
    PROCEDURAL FOR ac,dc =>
    vzeroa := (Vdd.v + Vss.v) * 0.5;  -- virtual 0v of opamp;

    -- ** CMRR **
    vd := [vinp,vinn].v;
    vc := (vinp.v + vinn.v)*0.5 - vzeroa;
    vca := vc * comgain;
    vin := vd + vca;

    --********** Limiter 1 *****************
    inl1 := vin;
    lim1 := Vdd.v - vzeroa;
    vinscale := inl1 / lim1;
    IF vinscale > 100.0 THEN vinscale := 100.0;
    ELSif vinscale <-100.0 THEN vinscale := -100.0;
    END IF;
    tanh_vin_scale := th(vinscale);
    outl1 := lim1 * tanh_vin_scale;

    --********** Pole1 *********************
    inp1 := outl1;
    sinp1 := ddt(inp1);
    soutp1 := ddt(outp1);
    -- soutp1/p1 + outp1 == inp1; ** IMPLICIT EQUATION **

    --********** Pole2 *********************
    inp2 := outp1;
    sinp2 := ddt(inp2);
    soutp2 := ddt(outp2);
    -- soutp2/p2 + outp2 == inp2 + sinp2/z1; ** IMPLICIT EQUATION **

    --********** Gain Stage ****************
    outg := outp2*gain;

    --********** Limiter 2 *****************
    inl2 := outg;
    lim2 := Vdd.v - vzeroa - limdiff;
    vinscale := inl2 / lim2;
    IF vinscale > 100.0 THEN vinscale := 100.0;
    ELSif vinscale <-100.0 THEN vinscale := -100.0;
    END IF;
    tanh_vin_scale := th(vinscale);
    outl2 := lim2 * tanh_vin_scale;

    -- *******Shift the output of the gain to be within the power supply **
    inls := outl2;
    outls := inls + vzeroa;

    --********** Slew rate limiting **********
    insr := outls; --Remove because of timestep function
    outsr := insr;

    --********** output resistance ***********
    inro := outsr;
    io := -vout.i;              -- vo.i is the current into vo !!
    IF (inro > 0.0) THEN ro := rop; -- rop = resistance out positive
       ELSE ro := ron;           -- ron resistance out negative
    END IF;
    vout.v %= inro - (io * ro);    -- ** assign the output **

    -- ********** Power Supply current ************
    idc := (Vdd.v - vzeroa)/ rdc;
    iac := gmdd * inro;
    Vss.i %= idc + iac + io;
    Vdd.i %= idc + iac - io;

--************************************************************;
-- ********* TRANSIENT *********************************;
    PROCEDURAL FOR transient =>
    vzeroa := (Vdd.v + Vss.v) * 0.5;  -- virtual 0v of opamp;

    -- ** CMRR **
    vd := [vinp,vinn].v;
    vc := (vinp.v + vinn.v)*0.5 - vzeroa;
    vca := vc * comgain;
    vin := vd + vca;

    --********** Limiter 1 *****************
    inl1 := vin;
    lim1 := 0.1; --Vdd.v - vzeroa;
    vinscale := inl1 / lim1;
    IF vinscale > 100.0 THEN vinscale := 100.0;
    ELSif vinscale <-100.0 THEN vinscale := -100.0;
    END IF;
    tanh_vin_scale := th(vinscale);
    outl1 := lim1 * tanh_vin_scale;

    --********** Pole1 *********************
    inp1 := outl1;
    sinp1 := ddt(inp1);
    soutp1 := ddt(outp1);
    -- soutp1/p1 + outp1 == inp1; ** IMPLICIT EQUATION **

    --********** Pole2 *********************
    inp2 := outp1;
    sinp2 := ddt(inp2);
    soutp2 := ddt(outp2);
    -- soutp2/p2 + outp2 == inp2 + sinp2/z1; ** IMPLICIT EQUATION **

    --********** Gain Stage ****************
    outg := outp2*gain;

    --********** Limiter 2 *****************
    inl2 := outg;
    lim2 := Vdd.v - vzeroa - limdiff;
    vinscale := inl2 / lim2;
    IF vinscale > 100.0 THEN vinscale := 100.0;
    ELSif vinscale <-100.0 THEN vinscale := -100.0;
    END IF;
    tanh_vin_scale := th(vinscale);
    outl2 := lim2 * tanh_vin_scale;

    -- *******Shift the output of the gain to be within the power supply **
    inls := outl2;
    outls := inls + vzeroa;

    --********** Slew rate limiting **********
    insr := outls;
    msrate := -1.0 * srate;
    diff := insr - outsr;
    rate := diff/time_step;
    IF rate > srate THEN diff2 := srate * time_step;
    ELSIF rate < msrate THEN diff2 := msrate *time_step;
    ELSE diff2 := diff;
    END IF;
    outsr := outsr + diff2 ;

    --********** output resistance ***********
    inro := outsr;
    io := -vout.i;              -- vo.i is the current into vo !!
    IF (inro > 0.0) THEN ro := rop; -- rop = resistance out positive
       ELSE ro := ron;           -- ron resistance out negative
    END IF;
    vout.v %= inro - (io * ro);    -- ** assign the output **

    -- ********** Power Supply current ************
    idc := (Vdd.v - vzeroa)/ rdc;
    iac := gmdd * inro;
    Vss.i %= idc + iac + io;
    Vdd.i %= idc + iac - io;

--************************************************************;
    EQUATION (outp1,outp2) FOR ac, dc, transient =>

    --** Pole1 **
    --  soutp1/p1 + outp1 == inp1;
        soutp1*ip1 + outp1 == inp1;

    --** Pole2 + Zero **
    soutp2*ip2 + outp2 == inp2 + sinp2*iz1;

    END RELATION;
END ARCHITECTURE a;
```

Fig. 4. VHDL-AMS code for open-loop op-amp.

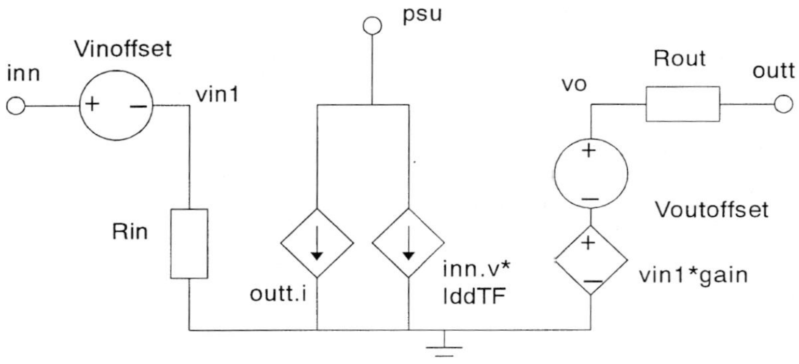

146 A. J. Perkins et al.

Fig. 5. Op-amp macromodel.

potentially large number of structural defects are assumed to manifest themselves as simple electrical faults. Even if the fault model is imprecise, test patterns designed to detect such faults will sometimes uncover other defects [8]. Analog circuits do not lend themselves to such simple fault models, not least because by definition, the range of possible behaviours is continuous rather than discrete. Nevertheless, by intuition it is likely that some defects will cause similar faulty behavior to others. If this can be shown to be true, the number of faults to be simulated would be reduced and hence the complexity

```
ENTITY opamp IS
   GENERIC(Rin, Vinoffset, Rout, gain,
           Voutoffset, IddTF : analog);
   PIN(inn, outt, psu : electrical);
END ENTITY opamp;

ARCHITECTURE c_loop OF opamp IS
   VARIABLE vin1, vo :analog;

BEGIN
   RELATION
     PROCEDURAL FOR init =>
       Rin := 400.0e3;
       Vinoffset := 40.0e-6;
       Rout := 1.0e3;
       gain := -50.0;
       Voutoffset :=0.0;
       IddTF := 70.0e-6;

     PROCEDURAL FOR ac,dc,transient =>
       vin1 := inn.v - Vinoffset;
       inn.i %= vin1 / Rin;
       vo := vin1 * gain + Voutoffset;
       outt.v %= vo - (-outt.i * Rout);
       psu.i %= -( (inn.v*IddTF) + outt.i);

   END RELATION;
END ARCHITECTURE c_loop;
```

Fig. 6. VHDL-AMS closed-loop op-amp model, including fault modeling.

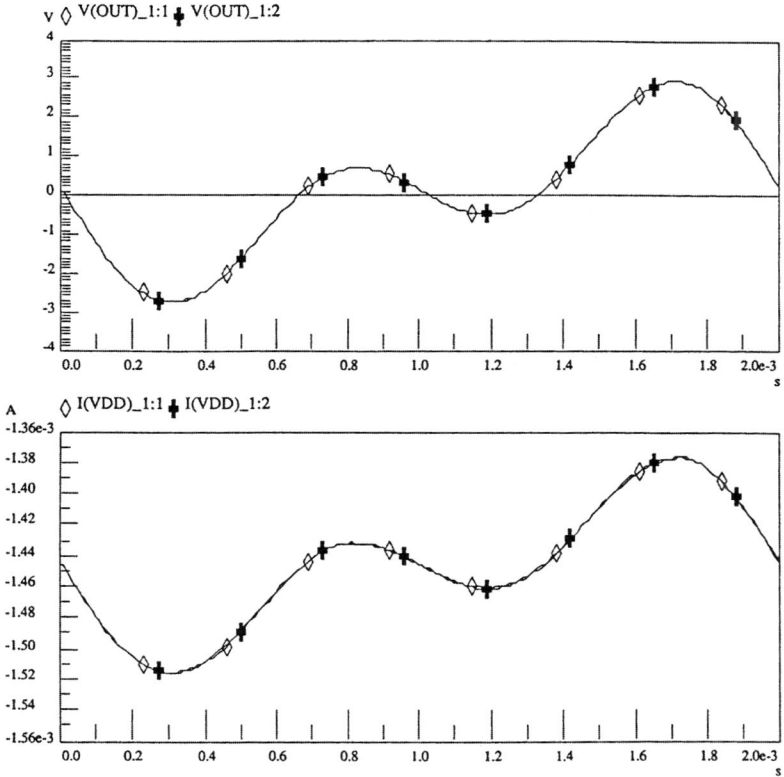

Fig. 7. A comparison between the VHDL-AMS and SPICE macromodel.

of fault simulation for analog circuits would be reduced.

A deeper problem exists concerning what constitutes a fault in an analog integrated circuit. Defects in the form of missing or extra metal or other material will cause bridging faults, open circuits and parametric changes to devices. In theory, the number of such potential defects over an integrated circuit tends towards infinity. Techniques and simulators [9] exist to predict the most likely defects, but the existence or otherwise of defects is dependent on the layout of the circuit. It is not therefore possible to derive a definitive fault list simply from the circuit schematic. As an alternative, it is possible to assume, for a small circuit, that a possible bridging fault exists between each network node and every other node. We discount, for the moment, bridging faults involving three or more nodes. Similarly, it is possible to assume that each branch of a circuit may become an open circuit. Finally, we can change the parameters of each network component. Of these possible faults, short circuits are probably the most likely, and can be modeled in SPICE as small resistances of e.g. 1 Ω between nodes [10]. Open circuits involving MOSFETs cause the drain source current to be zero at all times and can be modeled by setting the threshold voltage, V_T, to, say, 100 V [11]. Thus one fault model covers three faults. Both the fault models can be introduced without topological changes to the SPICE netlist, but introducing such changes manually is nevertheless a tedious task. Automatic generation of netlists containing faults and their automatic simulation is possible [12], but the time-consuming nature of analog fault simulation remains. The two-stage CMOS op-amp of Fig. 7 has 9 nodes, including the supply rails and 15 branches. There are, therefore 36 possible bridging faults involving two nodes (although we would normally choose to omit the case where the supplies are bridged) and 15 possible open circuit faults, giving a total of 51 possible faults and hence 52 simulations of such a circuit would be required. It will be appreciated that any technique to

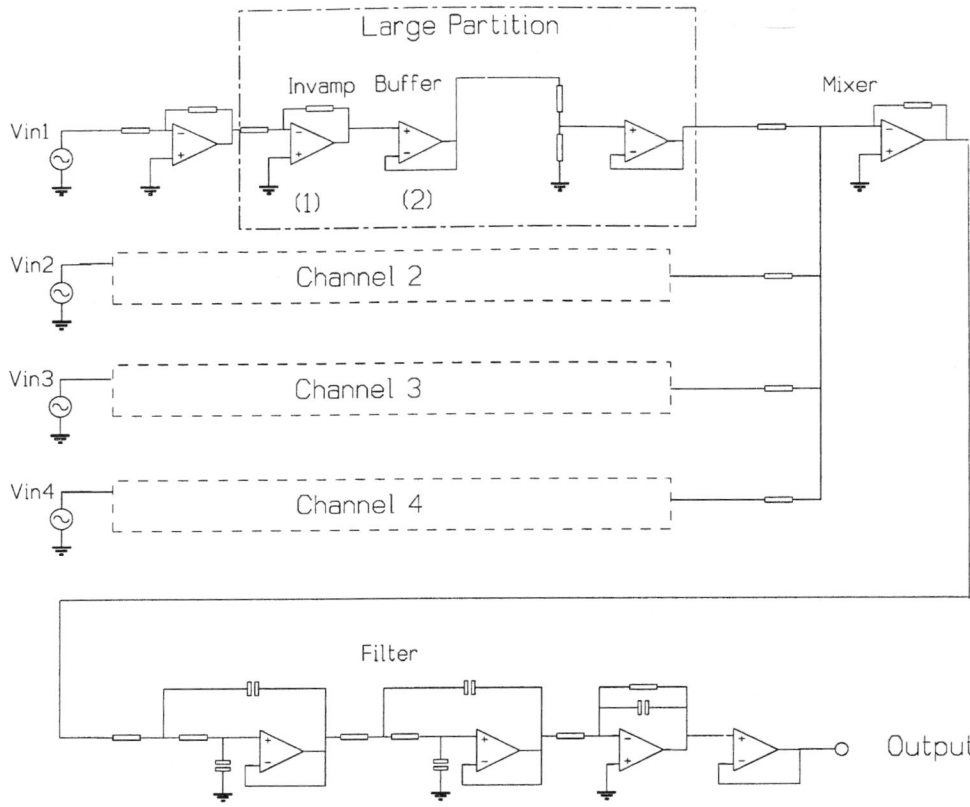

Fig. 8. Audio mixer circuit.

reduce the number of simulations is therefore of interest. Parametric changes have been omitted from consideration, not least because the distinction between a "fault" and a performance change due to parameter variation is difficult to define.

As the existence, or otherwise, of a fault is only observable at the outputs of a circuit, we have not considered the equivalence of faults in op-amps in open-loop configuration. Instead, we have again examined op-amps configured as inverting, non-inverting and summing amplifiers. The behavior of each circuit configuration was assessed using transient, AC and pole-zero analyses. The following measurements were taken:

- the output voltage offset (transient analysis with 100 Hz sine wave input)
- the RMS supply current (transient analysis with 100 Hz sine wave input)
- the voltage gain (v(out)/v(in)) (AC analysis)
- the supply current to input voltage admittance (i(vdd)/v(in)) (AC analysis)
- the input resistance (AC analysis)
- the output resistance (AC analysis)
- the voltage gain pole (AC & pole-zero analysis)

The two transient analyses can effectively be considered as DC sweeps. Table 2 shows how these measurements compare for a number of faults. If the all the measured values for a group of faults are the same or very similar, as shown, those faults can be collapsed, i.e. only one of the group needs to be simulated.

Certain faults cause difficulties in the SPICE analyses. For instance, a fault may cause the network matrix to tend towards singularity. At the very least, the simulation time increases; at worst the simulation fails completely. Other faults move the circuit away from the normal operating point. These latter faults will cause the output voltage to behave in a very non-

Fault Modeling and Simulation Using VHDL-AMS   149

Table 2. Selected results of fault simulations of the inverting amplifier.

Fault	vout	psu rms	vout tf	idd tf	Rout (out)	Rin	Pole (out)	Fault Group
fault free	0v	4e-6	−49.8	70u	6.40	400k	20k	
2 to 3	−5v	0.0	73u	0	1.5k	20M	391k	A
2 to 4	4.9	0	−13m	0	580	20.2M	2.6M	B
2 to 5	4.9	0	−10m	0	480	8M	85k	B
2 to 6	−5	0	74u	0	1.5k	20.4M	155k	A
2 to 7	4.9	0	29u	0	604	20.4M	100k	B
2 to 21	−5	0	74u	0	1.5k	20.4M	126k	A
3 to 6	−5	7.6e−7	74u	13u	1.5k	20.4M	1M	A
3 to 7	3.1	0	2	2.8u	35	19M	220k	D
3 to 21	4.9	0	−12m	0	570	20.1M	619k	B
4 to 21	−5	0	74u	0	1.5k	20.4M	4.2M	A
m1 o/c	4.94	0	18u	50n	375	20.4M	1.4M	C
m2 o/c	−5	0	74u	65n	1.5k	20.4M	5M	A
m3 o/c	−5	0	74u	0	1.5k	20.4M	1.5M	A
m4 o/c	4.91	0	−12m	0	577	20.4M	679k	C
m5 o/c	−4.84	0	117u	0	2.4k	20.4M	1M	A
m6 o/c	−5	0	74u	65n	1.5k	20.4M	772k	A

Table 3. Summary of fault collapsing.

	Bridging faults	No. open faults	Fault groups	Ungrouped	Unsimulatable	Nonlinear	Indistinguishable
Inverting	38	10	16	5			
Summing	38	10	12	15	1		
Non-inverting	32	10	6		1	14	1

linear way with respect to the input voltage. It is not generally possible to group such faults. Table 3 summarizes the fault collapsing for the three op-amp configurations. The number of fault simulations can be reduced in each case by up to 56%.

4. Behavioral Fault Modeling

For those fault groups that change the characteristics of the op-amp, while maintaining linear behavior, the faulty behavior can be mapped onto the SPICE behavioural model of the configured op-amp, by changing model parameters as specified in Table 2. Those faults that produce non-linear behavior require a more complex behavioral model. The linear controlled sources must be replaced by piecewise linear controlled sources. These models are not, however, available in all versions of SPICE. Exactly the same parametric changes may be made to the VHDL-AMS models to give behavioral models of faulty circuit blocks. The non-linear effects of certain faults can be modeled explicitly in VHDL-AMS.

In principle, therefore, simulations of complex analog (and mixed-signal) circuits can be performed in which fault-free and faulty circuit blocks are modeled behaviorally. Because of fault-collapsing, the number of fault simulations is reduced, and because of behavioral modeling, the complexity of the model and hence the simulation time is, in theory, reduced.

We tested this assumption with two circuits with multiple op-amps. The first circuit, an audio mixer shown in Fig. 8, has four identical input channels with four op-amps in each, followed by a mixing and filtering section with five op-amps. The circuit can be thought of as simply a chain of standard op-amp configurations. The second circuit is a leapfrog filter, Fig. 9, that has six op-amps and both local and global feedback. In both cases, we introduced faults into an op-amp and observed, by simulation at transistor and behavioral level, the effect on both the output and the supply current for the entire circuit. The detailed

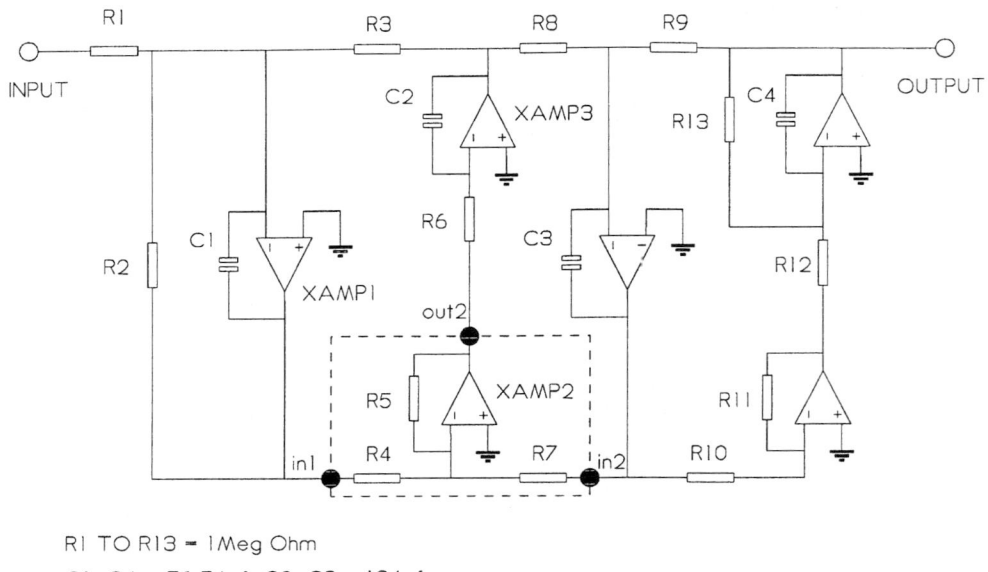

Fig. 9. Leapfrog filter.

results of these investigations are described elsewhere, but to illustrate the potential accuarcy of behavioral modeling, Figs. 10 and 11 show that the transistor models, the SPICE macromodels and the VHDL-AMS models give almost exactly the same voltage output and supply current behavior when faults are inserted into op-amp (1) in the mixer circuit. Fig. 10 shows the response to a fault in fault class A, while Fig. 11 shows the response to a fault from fault class B.

Certain faults, however, cause particular difficulties. In general, we consider a fault is detectable if it causes the supply current to vary by more than, say, 10%. Many of the behavioral models give an RMS supply current value with an accuracy to within 2% of that of the transistor models. As we are measuring the supply current of the entire circuit, in an attempt to emulate realistic test conditions, it should be noted that the introduction of a fault model may cause the output of that circuit block to take an abnormal value, and hence the supply current of succeeding stages may also differ from the fault-free value. The fault masking effect that has been observed with quiescent supply current monitoring in digital circuits [13] does not necessarily apply to analog circuits. Faults that cause the supply current to differ significantly from the fault-free case may cause the absolute value of the supply current simulated with behavioral models to be somewhat inaccurate when compared with the transistor model. It may be argued that these differences do not matter—provided that the absolute value is outside the normal circuit tolerances, its precise value is unimportant.

Further difficulties arise when the effect of a fault is to change either the input or output current or voltage of a stage to such an extent that the neighboring stage is operated outside its specified range. This is a particular, philosophical difficulty with all fault modeling. Models are in general designed to capture a particular set of characteristics. The introduction of a fault causes, by definition, abnormal effects. Therefore, if a fault causes a neighboring stage to operate in an unforeseen manner, the simulation results are likely to be inaccurate. This has been observed in the mixer circuit, where a bridging fault at the input of the buffer op-amp, (2) in Fig. 8, causes the previous stage to be heavily loaded. Similarly, in the leapfrog filter, the global feedback tends to reinforce fault effects in the highlighted op-amp, driving the output of that stage to one of the supply rails. This, in turn, takes the succeeding stage outside its normal operating range and causes it to saturate too. As we are concerned with modelling the supply current, it is important that these saturation effects should be

Fault Class A

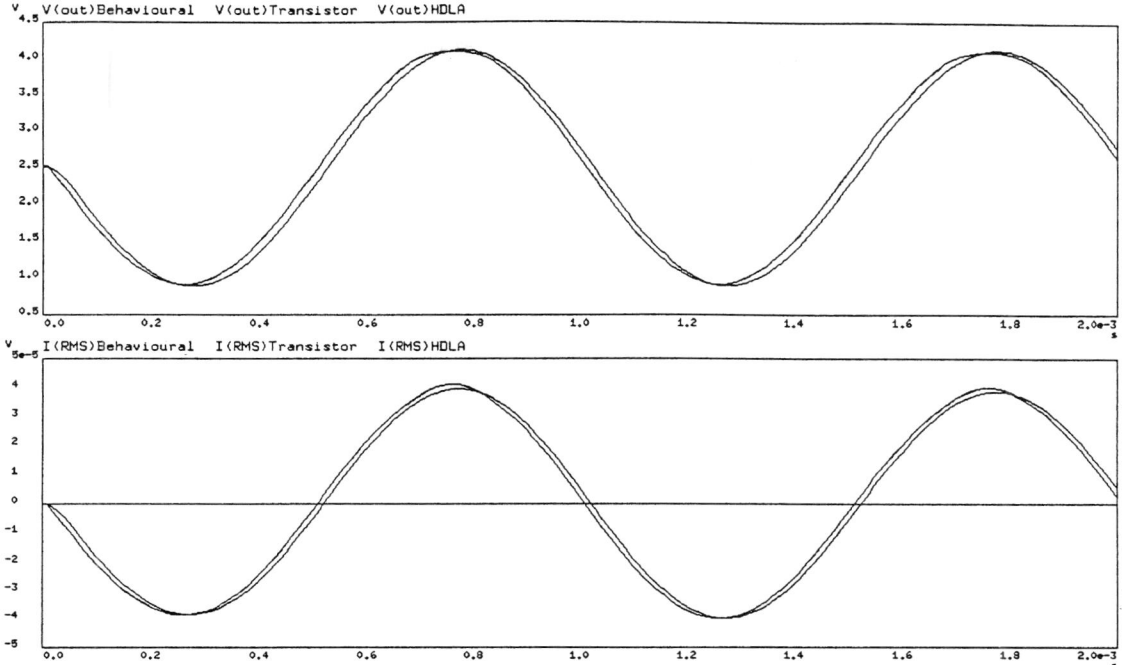

Fig. 10. Fault class A comparison between the transistor model, SPICE macromodel and the VHDL-AMS model.

properly modeled. In the leapfrog filter, the integrating stages are modeled using the open-loop op-amp model with a capacitor in the feedback loop.

These fault conditions can be modeled, behaviorally, in a number of ways. In the case of the mixer circuit, the scope of each behavioral block can be extended to embrace not one op-amp, but three. In effect therefore, we perform a higher level of fault collapsing and modeling. The effects of a fault are buffered from neighboring stages by implicit fault-free models of the adjacent op-amps, operating under faulty conditions. The fault thus occurs in the middle of a sliding window, which moves around the circuit according to the fault to be modeled.

The out-of-specification effects in the leapfrog filter are harder to model. The basic problem is that the behavioral model of an op-amp does not deliver enough current when the output is saturated. Thus the output stage of the macromodel can be enhanced, either by using more complex controlled sources, or even with a simple MOS transistor representation of the output stage.

The CPU times for simulations of two channels of the audio mixer of Fig. 8 for 2 ms with a stimulus of 1 kHz are shown in Table 4. Three modeling approaches are compared: a full transistor model; a SPICE behavioral model and a VHDL-AMS behavioral model. Three fault simulation times are shown, together with that of the fault-free simulation. It will be seen that the SPICE behavioral simulation is about 3 times faster than the transistor-level simulation, but that the VHDL-AMS simulation is significantly slower than both. Some reasons for this are discussed below.

5. Discussion

It has been noted that the use of an analog hardware description language for modeling both fault-free and faulty behavior of analog blocks is easier than the alternative: SPICE macromodeling. This is because of the greater flexibility of an HDL compared with the fixed structure of controlled sources and other circuit elements. This strength is, however, also a weakness. The incorporation of a semiconductor model into a

Fault Class B

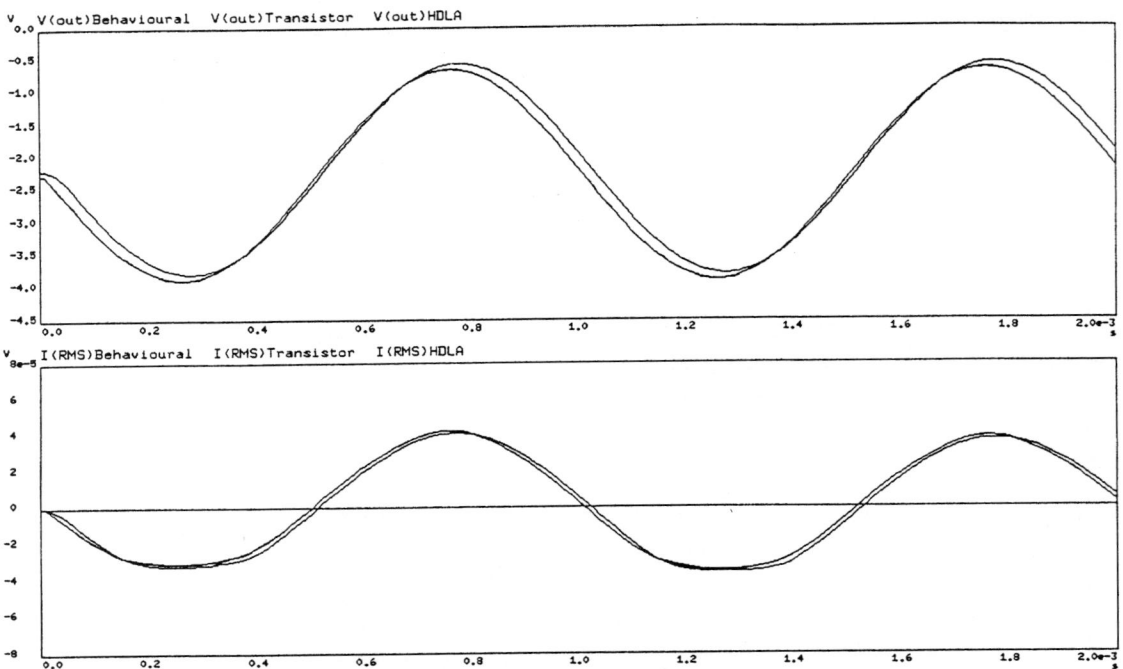

Fig. 11. Fault class B comparison between the transistor model, SPICE macromodel and the VHDL-AMS model.

circuit simulator, such as SPICE, is not a trivial task. Device equations must be formulated. These must be continuous throughout the expected range of behavior of the device. Moreover, it may be possible, during a Newton–Raphson iteration, for a circuit variable to, momentarily, take an unrealistic value. The model must be robust enough to handle this situation. Furthermore, because the Newton–Raphson algorithm uses the partial derivatives of a function with respect to all the controlling variables, these partial derivatives must be implemented and must also be continuous [14]. While VHDL-AMS does not require the user to specify partial derivatives, the other comments regarding model implementation apply.

Many of these difficulties become particularly apparent when models are made to operate in ways that were not originally envisaged.

Our experiences with VHDL-AMS have been gained through use of an early VHDL-AMS implementation—the HDL-A modeling language in the ELDO simulator from Anacad. We do not believe, however, that the difficulties we have encountered are necessarily a consequence of one particular simulator, but are likely to be endemic to all analog hardware description languages. The difficulties noted above with device models have also become apparent with VHDL-AMS descriptions. For instance, although it is possible to describe the transfer characteristic of an

Table 4. Comparison of CPU times for different modeling approaches.

	Fault-free	Fault Class A	Fault Class B	Fault Class L
Transistor level	39.0	34.0	35.9	37.3
SPICE macromodel	12.0	12.0	11.9	11.9
VHDL-AMS	55.0	53.0	55.0	57.0

op-amp as a piecewise linear function, it has been found that simply clamping a voltage causes timestep control difficulties.

This was evident in the simulation of the open loop op-amp of Fig. 3. Initially a piecewise linear voltage clamp was used, but this caused numerical oscillation when the input to the clamp changed rapidly. This often occurred when the op-amp, in a closed loop configuration, started to limit the output. As this occurred, the voltage difference between the input terminals would rise rapidly as the feedback no longer kept the negative terminal within a few microvolts of the positive terminal. The limiter was then changed to a hyperbolic cotangent function to smooth the transition from one region to another, which improved the stability, although the problem was not solved completely.

The use of VHDL-AMS was not found to deliver fully all of the desired benefits, namely quicker simulation speed and simpler representations of analog functional blocks. When comparisons between the SPICE and VHDL-AMS macro models of the fault model op-amp were done the VHDL-AMS actually simulated slower. This is almost certainly because the implementation of the simulator used was not fully optimized for VHDL-AMS as it included a full SPICE-compatible simulator as well. The VHDL-AMS parts are simulated using a secant method, which while easier to implement, converges more slowly to a solution at each time step [15]. Similarly, transitions between piecewise linear segments are not limited, which should prevent oscillations. To achieve a major decrease in simulation times the VHDL-AMS model would have to represent a lager circuit and hence a higher level of abstraction that was used here.

Conclusions

Fault simulation of analog circuits is important for confidence in analog and mixed-signal integrated circuits. At present, such fault simulation is of limited use, because of the speed of analog simulation and the large number of faults to be simulated. Simulation can be speeded-up by using simpler, behavioral models. The number of simulations can be reduced by fault collapsing. We have demonstrated that both these techniques yield valuable gains. Nevertheless, SPICE macromodeling is difficult. Analog hardware description languages offer a means to simplify behavioral modeling. The standard for VHDL-AMS is likely to be finalized in 1997, but it is apparent, from our investigations that behavioral modeling in general, and fault modeling in particular can cause difficulties for the underlying simulators. It is not unreasonable to suppose that the robustness and efficiency of VHDL-AMS simulators will improve, but it is still apparent that analog fault simulation *algorithms* need to be developed to support such modeling.

Acknowledgment

This work has been supported by EPSRC grants GR/K32661 and GR/J84120.

Appendix. Derivation of VHDL-AMS Op-amp model

The basic transfer function of the op-amp in Fig. 4 is given by:

$$\frac{V_{out}(S)}{V_{in}(S)} = \frac{(1 - s/z_1)}{(1 + s/p_1)(1 + s/p_2)}$$

Where p_1 and p_2 are the poles and z_1 is the high frequency zero. It is not possible to write the function in s notation directly, as VHDL-AMS only supports differential expressions. Rearranging the above expression, we get:

$$V_{out}(s)\left(1 + \frac{s}{p_1}\right) = V_{in}(s)\left(\frac{1 - s/z_1}{1 + s/p_2}\right)$$
$$= V_{p1}(s)$$

which can be rewritten as:

$$V_{out}(s) + \frac{1}{p_1} \cdot s \cdot V_{out}(s) = V_{p1}(s)$$

$$V_{p1}(s) + \frac{1}{p_2} \cdot s \cdot V_{p1}(s) = V_{in}(s)$$
$$- \frac{1}{z_1} \cdot s \cdot V_{in}(s)$$

The inverse Laplace transform of which is:

$$V_{out}(t) + \frac{1}{p_1} \cdot \frac{dV_{out}(t)}{dt} = V_{p1}(t)$$

$$V_{p1}(t) + \frac{1}{p_2} \cdot \frac{dV_{p1}(t)}{dt} = V_{in}(t) - \frac{1}{z_1} \cdot \frac{dV_{in}(t)}{dt}$$

The `ddt()` "function" is used to calculate the time derivatives of state variables (quantities), which are saved as new state variables. (Strictly, this is an implicit quantity, according to the proposed 1076.1 standard, and should be written as `v'dot`, not as a function.) Each of these expressions is now stated as an implicit equation. The remainder of the model is concerned with modeling saturation and other non-linear effects.

References

1. N. H. E. Weste and K. Eshraghian. *Principles of CMOS VLSI Design, A Systems Approach, 2nd Ed.* Addison-Wesley, Reading, MA, 1993, Chapter 7.
2. M. Abramovici, M. A. Breuer, and A. D. Friedman. *Digital systems testing and testable design.* Computer Science Press, New York, NY, 1990.
3. C. Di and J. Jess. "On accurate modeling and efficient simulation of CMOS opens." *Proc. IEEE Int. Test Conf.* pp. 875–882, 1993.
4. K. G. Nichols, T. J. Kazmierski, M. Zwolinski, and A. D. Brown. "Overview of SPICE-like circuit simulation algorithms.", *IEE Proc.-Circ. Dev. Syst.* 141, pp. 242–250, 1994.
5. D. Rodriguez. "Analog-VHDL—as an application, a real example." *IFIP Trans. A-Comp. Sci. and Tech.* 32, pp. 587–604, 1993.
6. G. A. Boyle, D. O. Pederson, B. M. Cohn and J. E. Solomon. "Macromodeling of Integrated Circuit Operational Amplifiers." *IEEE J. of Solid State Circuits* SC-9, pp. 353–363, 1974.
7. C. Chalk and M. Zwolinski. "Macromodel of CMOS operational amplifier including supply current variation." *Electronics Letters* 31, pp. 1398–1400, 1995.
8. T. M. Storey and W. Maly. "CMOS Bridging Fault Detection." *Proc. Int Test Conf.* pp. 842–851, 1990.
9. R. J. A. Harvey, A. M. D. Richardson, E. M. J. G. Bruls, and K. Baker. "Analogue Fault Simulation Based on Layout Dependent Fault models." *Proc. Int. Test Conf.* pp. 641–649, 1994.
10. A. Meixner and W. Maly. "Fault Modelling for the Testing of Mixed Signal Circuits." *Proc. Int. Test Conf.* pp. 564–572, 1991.
11. P. Caunegre and C. Abraham. "Achieving Simulation-Based Test Program Verification and Fault Simulation Capabilities for Mixed-Signal Systems." *IEEE ED&TC*, pp. 469–477, 1995.
12. S. J. Spinks and I. M. Bell. "Analogue Fault Simulation." *IEE Colloquium, Mixed Mode Modelling and Simulation*, No. 1994/205, 1994.
13. Y. K. Malaiya and A. P. Jayasumana. "Enhancement of resoltion in supply current based testing for large ICs." *Proc. IEEE VLSI Test Symp.* pp. 291–296, 1991.
14. T. L. Quarles. "Analysis of Performance and Convergence Issues for Circuit Simulation." Electronics Research Laboratory, University of California, Berkeley, CA, Memo. UCB/ERL M89/4, Chapter 3.
15. L. W. Nagel. "SPICE2: A computer program to simulate semiconductor circuits." Electronics Research Laboratory, University of California, Berkeley, CA, Memo. UCB/ERL M520, Chapter 5.

Brian Wilkins qualified in Electrical Engineering from University College, London. He joined Southampton University in 1965, and is now Senior Lecturer in charge of the Test Engineering Laboratory. For the past 15 years he has been working on various problems associated with testing and DFT, most recent on implementation of boundary scan, testability guidelines, and analog and mixed-signal testing. He is a member of the IEEE Computer Society and a Fellow of the IEE.

Christopher Chalk received his B.Sc. (physics with electronics) from the University of East Anglia (UK) in 1986. In 1994 he received an M.Sc. in electronics from the University of Southampton (UK). He is currently working as a research assistant at the University of Southampton in collaboration with the Universities of Hull and Huddersfield evaluating novel mixed signal test methodologies in addition to studying for a Ph.D. on AC RMS power supply current monitoring. Prior to this, he has held positions at the

British Broadcasting Corporation as a Radio Test Engineer. His current research interests include power supply current monitoring, automatic test pattern generation, behavioral modeling with high level description languages (VHDL-AMS and spectreHDL), built in current sensors and design for testability.

Andrew Perkins is a research assistant in the Electronics and Computer Science Department at the University of Southampton. His research interests include the design and test of analog and mixed signal integrated circuits. He received a B.Eng.(Hons) degree from Manchester Polytechnic and an M.Sc. from the University of Southampton.

Mark Zwolinski gained his B.Sc. and Ph.D. from the University of Southampton. He has been a lecturer in the Department of Electronics and Computer Science at the University of Southampton since 1990. His research interests include mixed-signal simulation, mixed-signal test and synthesis. He has published over 30 papers in the fields of design automation and test and is co-author of a book on circuit simulation. He is a member of the IEE, the IEEE, ACM and is a Chartered Engineer. He is also a member of the IEEE Computer Society Test Technology Committee and the ACM Special Interest Group on Design Automation.

Creative Methods of Leveraging *VHDL-AMS*-like Analog-HDL Environments. Case Study: Simulation of Circuit Reliability

SERAG M. GADELRAB* AND JAMES A. BARBY**

Nortel, Ottawa, Ontario, Canada
**Department of Electrical and Computer Engineering, University of Waterloo, Waterloo, Ontario, Canada N2L 3GI*

Received August 1, 1996; Accepted August 20, 1997

Abstract. Given an Analog Hardware Description Language (Analog-HDL), like the proposed VHDL-AMS, one can do much more than conventional component modeling. One can develop additional simulation capabilities by creatively using the Analog-HDL/VHDL-AMS environment. To demonstrate this, we present an innovative method of implementing simulation of systems whose equations (or parameters) change with time (such as reliability simulation, component failure modeling and sensitivity analysis). The simulation algorithm is defined in terms of event-driven control modules, signals and interfaces within a generic Analog-HDL environment. To verify our methodology, we use this configuration to implement three different circuit reliability simulation algorithms. We describe the different modules required to implement the algorithm in terms of IEEE 1076.1 PseudoCode routines. Our implementation allows the definition of two concurrent versions of time within the same simulation. We present a comparison of the results obtained from applying the three reliability simulation algorithms to amorphous silicon thin-film transistor circuits.

Key Words: VHDL-AMS modeling environment, simulation and modeling methodology, user-extended simulation algorithms, time varying models, reliability simulation, failure effects modeling analysis

I. Introduction

The process of creating special purpose simulators that accommodate newly defined simulation algorithms can be a time consuming effort that encompasses the development of simulator engines, simulator pre-processors and post-processors, and user graphic interfaces. The time invested in the development of such new simulation algorithms may be greatly reduced if they are implemented within an Analog-HDL simulation environment. We define such an environment as one which allows the simulation of user-defined models of physical systems using user-extended simulation algorithms that sit on top of a basic analog simulation engine and a discrete event simulation engine. The first-generation of SPICE-like simulation environments, which were targeted at the design of small analog circuits, did not provide such capabilities. Recently developed Analog Hardware Description Language environments[1] offer a framework for the development and implementation of arbitrary components and/or system models over a wide range of abstraction spanning from traditional, SPICE-like, component models to higher-level behavioral models [1–3]. However, to our knowledge, the ability of Analog-HDLs to support user-extended simulation algorithms has yet to be investigated.

The primary goal of this work is to show that generic Analog Hardware Description Languages and the simulation engines that support them are capable of supporting user-extended simulation algorithms. With the aid of Fig. 1, we will show that generic Analog-HDL environments (like VHDL-AMS) possess all the pieces that are required to define and implement arbitrary simulation algorithms. The conventional configuration of an Analog-HDL environment, which is shown in Fig. 1(a), may be divided into two domains: the user-defined and the vendor-defined domains. The user-defined domain is composed of a netlist that describes the physical system under simulation. This netlist represents interconnected instances of user-defined software structures that model physical components. In this work, we refer to these software structures as *modules* (or entity-architecture pairs in VHDL-AMS). The vendor-defined domain contains a collection of

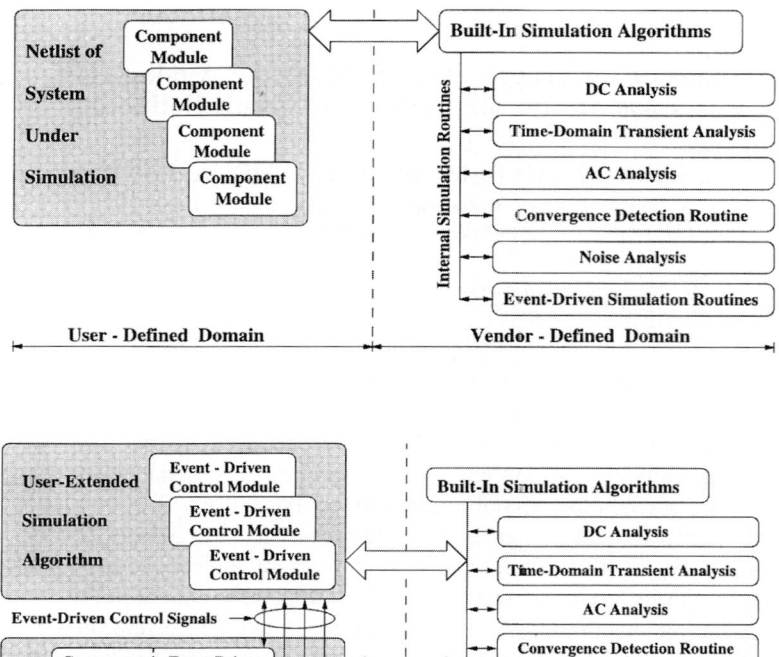

Fig. 1(a). The conventional configuration of an Analog-HDL environment. *Fig. 1(b).* The proposed configuration for implementing arbitrary simulation algorithms within an Analog-HDL environment.

built-in simulation algorithms and internal simulation routines (such as transient and event-driven simulation routines). The user creates the netlist (including the modules that represent component models), then selects a simulation algorithm from the limited number of built-in simulation algorithms. Built-in simulation algorithms may use one or more internal simulation routines. For example, a mixed-signal time-domain transient simulation algorithm will require the use of three routines: DC analysis, time-domain transient analysis and the event-driven simulation routines.

In order to implement user-extended simulation algorithms within a generic Analog-HDL environment we propose the configuration shown in Fig. 1(b).

In this configuration, a user-extended simulation algorithm is specified in terms of a set of user-defined, **event-driven control (EDC)** modules. These modules exist in the user-defined domain rather than the vendor-defined domain (see Fig. 1(b)). The EDC modules implement the user-extended simulation algorithm by influencing the behavior of the system that is under simulation. This implies the presence of a communication link between the EDC modules and the system that is under simulation. Such a link can be realized through a set of event-driven control signals as shown in Fig. 1(b). The system that is under simulation must be equipped with the ability to process and respond to these event-driven control signals. This is achieved by adding a simple event-driven interface which is built into the component modules. It is important to note that our approach does not allow, nor expect, the simulation algorithm to be implemented by changing the built-in, vendor-defined, simulation algorithms or routines. Rather, our approach involves the *addition* of new simulation algorithms which utilize the simulation routines that already exist within the simulation environment.

In the following sections we demonstrate that the configuration shown in Fig. 1(b) can be used to realize user-extended simulation algorithms within generic Analog-HDL environments. Using circuit reliability simulation algorithms as an example, we describe practical implementations of EDC modules and event-driven interfaces that allow the creation of a circuit reliability simulation algorithm within a generic Analog-HDL environment. We begin, in Section II, by presenting the three circuit reliability simulation algorithms that are available in the literature along with a summary of how these algorithms are currently being implemented using SPICE-like simulators [4–8]. We then explain how these reliability simulation algorithms can be implemented within an Analog-HDL environment (like VHDL-AMS) using the framework shown in Fig. 1(b); this is done by describing the construction and workings of the event-driven interfaces and the event-driven control modules in Sections III and IV, respectively. Throughout these two sections, we provide IEEE 1076.1 PseudoCode routines of the key event-driven parts in order to demonstrate the operation of the different modules. In Section V, we present the results of applying the three reliability simulation algorithms to amorphous silicon thin-film transistor circuits. Finally, we discuss, in Section VI, other simulation algorithms that can easily be implemented in a generic Analog-HDL environment using our methodology.

II. Background to Circuit Reliability Simulation

Unlike time-domain transient simulations, a circuit reliability simulation spans two different temporal axes (see Fig. 2). These two temporal axes are *time* and *age*. All simulations conducted along the *time*

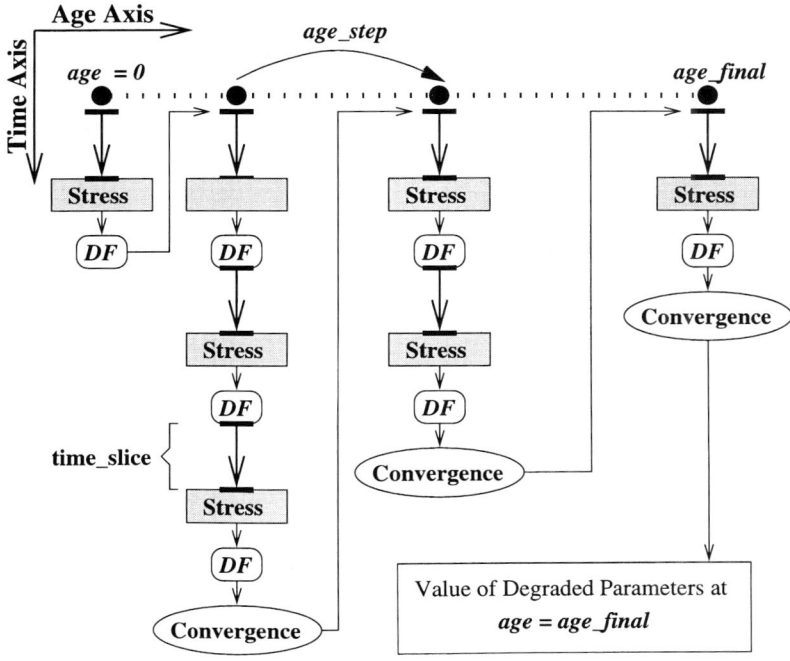

Fig. 2. The multi-step, iterative reliability simulation algorithm.

axis represent short temporal periods during which component degradation is negligible, for example, the period of a typical time-domain circuit transient simulation (fractions of a second). We call such short simulation periods **time_slice**s. The *age* axis represents *long* temporal periods during which component degradation is significant and can not be neglected (months or years). The incremental unit along the *age* axis is an *age_step*. A complete reliability simulation algorithm is shown in Fig. 2. The reliability simulation starts, at *age* = 0, by running a transient circuit simulation for the duration of one **time_slice** along the time axis. Once this transient simulation is completed, the **stresses** experienced by each device in the circuit at *age* = 0 are extracted from the transient voltage and/or current waveforms. The simulator substitutes the values of these stresses into a collection of **degradation functions**, denoted by *DF*, which predict the new, degraded values of device parameters at

$$age = current\_age + age\_step$$

However, since the amount of stress experienced by a given device is continuously changing throughout an *age_step*, the values of the degraded device parameters must be iterated until their values converge at a given *age*. Each iteration is performed by running a time_slice simulation followed by a re-evaluation of the degraded device parameters as seen in Fig. 2. Once convergence is achieved at the current value of *age*, the reliability simulation advances to the next value of *age*. The process is repeated until the final age, *age_final*, is reached. This reliability simulation algorithm is therefore a **multi-step, iterative algorithm**.

A multi-step, iterative reliability simulation algorithm was implemented by [4] using an external software shell in concert with HSPICE.[2] Initially, the external shell generates a circuit netlist for HSPICE and runs it for a period of one time_slice. Upon the completion of the time_slice, the external shell reads the output files generated by HSPICE, computes the amount of stress experienced by each device, calculates the amount of degradation in the parameters of each device, regenerates a circuit netlist, which now includes the effect of degradation, and then prompts HSPICE to read the revised netlist and conduct a new time_slice simulation. This external shell approach has many disadvantages. First, it requires the creation of an external shell. Second, it relies on a SPICE-like transient simulator to generate the value of device stresses. Most SPICE-like transient simulators come with a variety of problems such as the limited number of device models which they offer, the lack of an easy method for incorporating new device models into the simulator and the inability to simulate mixed-discipline systems like mechatronics. Finally, the external shell architecture is costly in terms of CPU time since it involves running a large number of sequential transient simulations, each requiring a new DC solution and a new updated netlist (which must be read and compiled by HSPICE). In order to reduce the CPU time associated with the external shell implementation, almost all external shell-based reliability simulators [5–7] abandon the multi-step, iterative simulation algorithm and adopt the less accurate, **single-step, non-iterative algorithm**. In this algorithm only one time_slice simulation is performed at *age = 0*. The stress values obtained from this time_slice are used to advance directly to the final age without any iterations.

Recently, Leblebici and Kang [8] attempted to overcome the disadvantages of the external shell-based simulators by developing an integrated reliability simulator using iSMILE [9]. The simulator uses a **multi-step, non-iterative algorithm**. The stresses associated with a given device parameter are monitored and computed by adding physical capacitors and current-sources to each device instance in the circuit. The **degradation functions** are also represented by physical capacitors and non-linear current-sources within the device instance. This results in a large number of additional circuit nodes[3] and, hence, an increase in the simulation time.

In the following two sections we discuss the implementation of the three circuit reliability simulation algorithms discussed above within an Analog-HDL environment (like VHDL-AMS) as described in Fig. 1(b). Where relevant, we also compare the Analog-HDL based implementation to the conventional, SPICE-based implementations.

III. The Event-Driven Interface

Reliability simulations are usually conducted on the transistor level and hence the component modules (shown in Fig. 1(a) and (b)) are electrical modules. In order to implement user-extended reliability simulation algorithms, we append an event-driven interface to these electrical modules in accordance with Fig. 1(b). Adding an event-driven interface to an electrical module transforms it into an **event-driven electrical (EDE) module** (or entity-architecture pair in VHDL-AMS). Fig. 3 shows an EDE module for a three terminal FET. Within the EDE module, the equations that describe the behavior of the device during a time-domain transient simulation (i.e. the DC and transient equations of the transistor) are contained in the **electrical component** portion. The EDE module is similar to the conventional electrical module in that it has only three electrical connections. Unlike the conventional electrical module, the EDE module has event-driven input and output pins which connect the event-driven interface to the event-driven control modules (refer to Fig. 1(b)). These pins are not electrical nodes but rather discrete **analog-states** (also know as **sampled-data** variables); that is, signals which are continuous in value but discrete in time (i.e. signals of type REAL in VHDL-AMS). The event-driven signal pins are <u>not</u> included in the system matrix and, hence, do not add to the computational complexity of the transient simulation. Therefore, within a time-domain transient simulation, the electrical and the event-driven electrical modules of a given device behave identically.

The **event-driven interface** of the EDE module contains the **degradation functions** of the device and administers the process of aging at the device module level. The functionality of the **event-driven interface** is described by the IEEE 1076.1 PseudoCode routine shown in Fig. 4. This code is based on extensions to the basic VHDL [10] capabilities. In the PseudoCode, the italics font is used to represent variables that are local to the process (e.g. *average_gate_voltage*) and the typewriter font is used to represent signals that are connected to the module via ports (e.g. `update`). The interface quantities of the module are in normal font (e.g. gate_voltage). The event-driven interface calculates the stress experienced by the EDE module during a time_slice simulation through an internal **wake-up** mechanism: at the end of each integration time-step (wait on `time_step_complete`;), an internal **wake-up** routine prompts the event-driven interface to compute the stress experienced during the past integration time-step and add this value to a variable that is local to the process (*average_gate_voltage*) that represents the total stress experienced by the individual device. The event-driven interface then returns to a dormant state until it is wakened by the completion of the next integration time-step. This event-driven process takes an insignificant amount of CPU time as it is a sequential series of addition operations. By using the event-driven interface, the Analog-HDL based simulator avoids the CPU expensive processes associated with computing the value of device stresses in external shell-based reliability simulators [4–7] and the iSMILE-based reliability simulator [8].

During the period of a time_slice, the event-driven input and output signals are inactive and, hence, do not affect the operation of the EDE module. However, once a time_slice simulation is completed, these event-driven input signals alert all event-driven interfaces (wait on `update`;) and notify each EDE module of the value of the `current_age`. This triggers a series of events within each event-driven interface. First, the final value of stress is updated. Then, the new values of the degraded device parameters are computed using the device **degradation function**.[4] These values are then loaded into the electrical component portion of the module such that the electrical model reflects the degradation incurred during the past *age_step*.

Having updated the value of the degraded parameters, the event-driven interface checks for convergence by computing the change incurred in

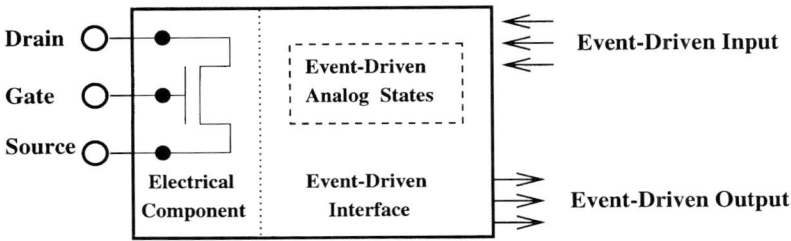

Fig. 3. An event-driven electrical module for a three terminal FET.

```
--********************************************************************
-- Event driven interface of electical model including degradation function
--********************************************************************
...
-- the following would be **generic constants** in the entity/architecture:
--            time_slice, tau, beta, initial_threshold_voltage, user_defined_tolerance
-- the following are **local variables** to the architecture:
--            age, average_gate_voltage, prev_threshold, current_threshold,
--            threshold_at_prev_age, age_step, delta_threshold
-- the following is an **interface quanitity** of the architecture:
--            gate_voltage
-- the following are **signals** that would be connected to the architecture via ports:
--            update, current_age, conv, time_step_complete
...
process begin
     if initialization_phase then
          age := 0.0;
          average_gate_voltage := 0.0;
          prev_threshold := initial_threshold_voltage;
          current_threshold := initial_threshold_voltage;
          threshold_at_prev_age := initial_threshold_voltage;
     end if;
     loop
          -- updating average gate stress at end of each time step
          -- time_step is a function
          wait on time_step_complete;
          average_gate_voltage := average_gate_voltage
                              + gate_voltage * time_step/time_slice;
     end loop;
end process;
-- Updating the threshold voltage at end of each time_slice
process begin
     wait on update;
     age_step := current_age - age;
     --
     -- the DEGRADATION FUNCTION
     current_threshold := average_gate_voltage - ( average_gate_voltage
                    - threshold_at_prev_age)* exp(-1.0*((age_step/tau)**beta));
     --
     average_gate_voltage := 0.0;
     --
     -- Check for convergence
     delta_threshold := abs(current_threshold - prev_threshold):
     if delta_threshold < user_defined_tolerance then
          conv <= '1';
          threshold_at_prev_age := current_threshold;
          else
               conv <= '0';
          end if;
          prev_threshold := current_threshold;
end process;
-- Updating the current age counter
process begin
     wait on current_age;
     age := current_age;
end process;
...
-- conventional electrical model
...
```

Fig. 4. The IEEE 1076.1 PseudoCode routine for the **event-driven interface**.

the value of each device parameter during the most recent time_slice. If the change is less than a user-specified tolerance, the event-driven interface declares this parameter to have converged. Each degrading parameter controls the digital state of an event-driven output convergence flag signal. Setting the flag value to 1 means that the parameter has converged while resetting the flag value to 0 means that it has not converged. The user-defined event-driven control modules (Fig. 1(b)) monitor the convergence state of the circuit through these convergence flags.

Referring to Fig. 2 we observe that the reliability simulation is a series of time_slices that are separated by the process of calculating the value of the degraded parameters for each device. At the end of each time_slice, the Analog-HDL environment updates the value of the degraded parameters within each EDE module, then proceeds to the next integration time-step which lies within the following time_slice. Therefore, the EDE module allows the merging of consecutive time_slices into one continuous event-driven transient simulation as shown in Fig. 5. Hence, the *time* temporal axis of the reliability simulation is represented by the intrinsic time parameter of the Analog-HDL environment. The second temporal axis, *age*, is realized through the event-driven control modules which are described in Section IV.

It is important to note that after a time_slice has been completed there is no need to save the dynamic state of the system for retrieval prior to starting the following time_slice. Rather, following the completion of the last integration point within a time_slice, the analog-solver automatically advances to the next integration point, which happens to occur within the following time_slice. In doing so, the analog-solver automatically uses the last state of the system (which is the last integration point in the previous time_slice) as the past state, and proceeds with calculating the new state of the system as it would for any time-step. As such, time is not reset at the end of each time-slice but rather is continued to be incremented since the analog-solver believes that it is progressing through a single analog simulation rather than two consecutive time_slices. This feature results in a considerable savings in terms of simulation time and disk-space since the simulator need not save (and then retrieve) the dynamic state of the system at the time_slice boundaries.

The Analog-HDL based implementation compares favorably with external shell-based reliability simulators [4–7] which terminates the time-domain transient simulation at the end of each time_slice in order to update the value of the degrading parameters. The procedure used by [4–7] may be costly in terms of CPU time since it involves the generation of a new netlist and the calculation of a new DC solution for each time_slice.

IV. The Event-Driven Control Modules

Fig. 6 shows the user-defined domain of an Analog-HDL environment when applied to a reliability simulation. In accordance with the methodology presented in Fig. 1(b), the simulation algorithm for the reliability simulation is administered by three **event-driven control (EDC)** modules: **age keeper**, **time slicer** and **convergence umpire**. Communication between the system that is under

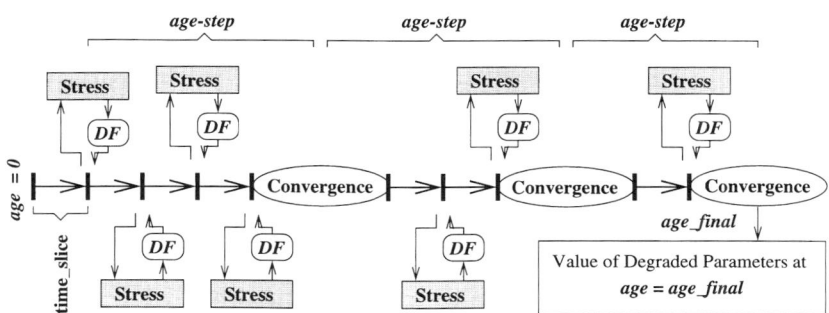

Fig. 5. The realization of the *age* axis by merging consecutive time_slices into a single event-driven transient simulation.

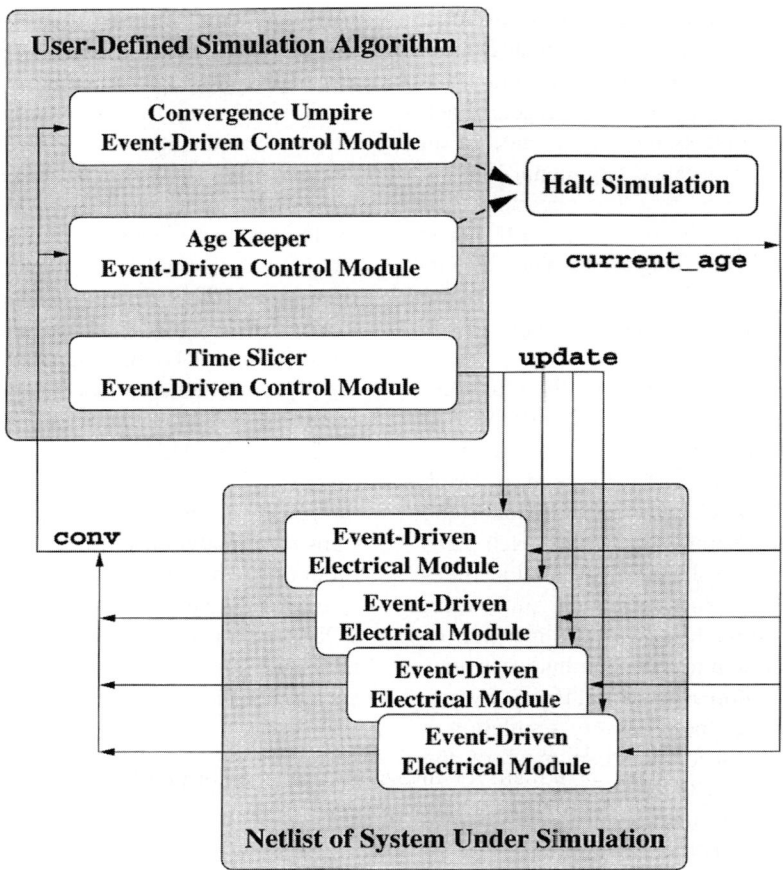

Fig. 6. The user-defined domain of an Analog-HDL environment for a circuit reliability simulation.

simulation and the three EDC modules is conducted through three event-driven signals: `conv` which is the global convergence flag, `update` which is the time_slice clock signal and `current_age` which is a signal that is discrete in time but continuous in value (i.e. a signal of type REAL) and has the task of conveying the current value of *age* to all EDE modules. The administration of the circuit reliability simulation algorithms can be explained by outlining the functions of the three EDC modules. This is done with the aid of the IEEE 1076.1 PseudoCode routines for these modules which are shown in Fig. 7.

The primary task of the **time slicer** is to alert the event-driven interfaces of all EDE modules at the end of each time_slice. Using an internal wake-up routine, the **time slicer** schedules an event on the `update` signal at the end of every time_slice period. Such an event on `update` triggers the event-driven interfaces of all EDE modules causing them to update the values of the degrading parameters and to report the state of their convergence. Once the component parameters are updated, there may be an abrupt discontinuity in the electric behavior of the EDE modules. Therefore, the **time slicer** ensures that two integration time-steps are scheduled: one immediately before the `update` signal and the other immediately after the activation of the `update` signal. This assists the convergence of the integration routine given the abrupt discontinuity in the device behavior[5] [11].

The **convergence umpire** module detects convergence of the reliability simulation when an iterative algorithm is being used. In order to detect global convergence, all event-driven convergence output flags are connected together into a single global

```
-- *********************************************************************
-- time_slicer algorithm
-- *********************************************************************
...
-- the following would be **generic constants** in the entity/architecture:
--            time_slice, delta_time
-- the following would be **local signals** of the architecture:
--            wake_up1, wake_up2, time_step_flag
-- the following would be **signals** that are connected to the architecture via ports:
--            update
...
-- Scheduling Outputs for update
process begin
      if initialization_phase then
            update <= '0' after time_slice;
            -- Wake-up mechanism for update flag
            wake_up1 <= '0' after time_slice;
      end if;
      loop
            wait on wake_up1;
            update <= not(update) after time_slice;
            wake_up1 <= not(wake_up1) after time_slice;
      end loop;
end process;
-- Scheduling integration-steps before end of time_slice
process begin
      if initialization_phase then
            -- Wake-up mechanism for integration-step before end of time_slice
            wake_up2 <= '0' after time_slice-delta_time;
            -- Schedule integration-step before and after time_slice boundary
            time_step_flag <= transport '-1' after time_slice-delta_time,
                                         '+1' after time_slice+ delta_time;
      end if;
      loop
            wait on wake_up2;
            time_step_flag <= transport '-1' after time_slice,
                                         '+1' after time_slice+2.0* delta_time;
            wake_up2 <= not(wake_up2) after time_slice;
      end loop;
end process;
-- force an analog solution time step using the force_time_step procedure
process begin
      wait on time_step_flag;
      force_time_step;
end process;
-- end of time_slicer
```

Fig. 7(a). The IEEE 1076.1 PseudoCode routines for the **time slicer**.

signal, conv. The **convergence umpire** uses the arbitration rules for signals to check whether global convergence has occurred. If all convergence output pins are of the same level (1 or 0), then the simulation environment assigns that level to conv. Therefore, if all EDE modules converge, then conv=1 and if none of the EDE modules converge, then conv=0. If some EDE modules have converged while others have yet to converge, there will be a conflict on the state of conv. In this case, the simulation environment would

```
-- **************************************************************
-- Conv_umpire algorithm
-- **************************************************************
...
-- the following would be **generic constants** in the entity/architecture:
--          max_iteration
-- the following would be **local variables** of the architecture:
--          internal_counter
-- the following would be **signals** connected to the architecture via ports:
--          conv
...
-- Initialize internal counter
process begin
    if initialization_phase then  --schedule first age_step
        internal_counter := 0.0;
    end if;
    loop
        wait on conv;
        if conv /= '1' then
            internal_counter := internal_counter + 1.0;
            if internal_counter > max_iteration then
                report "Number of maximum iterations exceeded"
                severity FAILURE;
                halt_simulation(NOW);
            end if;
        end if;
    end loop;
end process;
```

Fig. 7(b). The IEEE 1076.1 Pseudo Code routines for **convergence umpire**.

assign the unknown state to conv. Therefore, global convergence occurs when, and only when, conv=1. This mechanism for identifying global convergence is also used by the **age keeper** module.

The **convergence umpire** also monitors the number of time_slices that have been simulated at the current_age point by counting the number of events that occur on conv. Since all EDE modules generate events on their convergence flags at the end of each time_slice, an event on the conv flag can be interpreted as the end of a time_slice. If the number of time_slices exceeds a user-specified maximum per *age_step*, the **convergence umpire** terminates the aging simulation and informs the user. When convergence is achieved at a given *age*, the **convergence umpire** resets its internal time_slice counter.

The **age keeper** module has the role of realizing the second temporal axis, *age*. This task consists of maintaining and updating the value of current_age in accordance with the convergence state of the degrading device parameters. During an iterative simulation, the **age keeper** module monitors the conv flag. Once convergence is achieved, it increments the value of its local variable that represent *age* according to a user supplied algorithm and then passes this value to all EDE modules via the current_age signal. Such increments along the *age* axis can be either fixed-value steps or adaptively computed steps whose values depend on the convergence history of the reliability simulation. Once convergence is achieved at *age_final*, the **age keeper** module terminates the reliability simulation.[6] During a non-iterative simulation, **age keeper** increments the

Creative Methods of Leveraging *VHDL-AMS*-like Analog-HDL Environments

```
--**********************************************************************
-- Age_keeper algorithm
--**********************************************************************
...
-- the following would be generic constants in the entity/architecture:
--           age_step, max_iteration, age_final
-- the following would be local variables of the architecture:
--           age
-- the following would be signals connected to the architecture via ports:
--           current_age, conv
...
process begin
    if initialization_phase then  --schedule first age_step
        age := age_step;
        current_age <= age;
    end if;
    loop  -- update mechanism & simulation termination condition
        wait on conv;
        if max_iteration = 1.0 then  -- multi-step, no-iteration routine
            age := age + age_step;
            if age > age_final then halt_simulation(NOW);
            else current_age <= age;
            end if;
        else  -- multi-step, with-iteration routine
            if conv = '1' then
                age := age + age_step;
                if age > age_final then halt_simulation(NOW);
                else current_age <= age;
                end if;
            end if;
        end if;
    end loop;
end process;
```

Fig. 7(c). The IEEE 1076.1 Pseudo Code routine for **age keeper**.

current_age at the end of each time_slice, regardless of the state of conv. The choice of simulating using iterative or non-iterative methods is made by the user via the appropriate input parameters for **age keeper**.

V. Simulation Results

To demonstrate the feasibility of implementing an arbitrary simulation algorithm within a generic Analog-HDL environment, we constructed a prototype reliability simulator using a commercially available Analog-HDL based environment, Saber [1,12]. We used this commercial simulation environment since the official IEEE 1076.1 environment is not presently available. However, we believe that if two Analog-HDL environments have the same modeling functionality, then one can map models from one Analog-HDL to another Analog-HDL. We generated the three EDC modules described in Section IV and an EDE module for amorphous silicon (a-Si:H) thin-film transistors (TFTs). Amorphous silicon TFTs suffer from bias-induced instability of the threshold voltage, V_{thr}. The following threshold voltage degradation function [13] was implemented in the event-driven interface of the transistor EDE module:

$$V_{thr} = V_{GS} - (V_{GS} - V_{thr}(t_0)) \times \exp^{-(t/\tau)^\beta} \quad (1)$$

Here $V_{thr}(t_0)$ is the threshold voltage at $age = 0$, V_{GS} is the gate-to-source voltage, β is the dispersion parameter $(0.25 \leq \beta \leq 0.65)$ and τ represents the density of weak bonds $(\tau = \tau_0 \times \exp^{(E_a/kT)}$, where $\tau_0 = 10^{-11}$ and $E_a = 1.1\ eV)$. The value of β was assumed to be 0.35. The total TFT degradation is calculated as a function of the weighted average of the gate-to-source voltage during a time_slice. This average value is computed inside the event-driven interface of the amorphous silicon TFT module according to the following equation:

$$V_{GS} = \frac{\sum_{i=1}^{N} V_{GS,i} \times t_{integration-step,i}}{\sum_{i=1}^{N} t_{integration-step,i}} \quad (2)$$

Where N is the number of integration-steps in a time_slice, i is the integration-step index, $t_{integration-step,i}$ is the period of the ith integration step and $V_{GS,i}$ is the gate-to-source voltage during the ith integration-step. The electrical model for the TFT uses level-1 MOS I–V equations. These equations give accurate results when used with a reduced mobility $(0.1 \leq \mu \leq 1.0\ cm^2/V.sec)$ [14]. The TFT capacitance was computed using Meyer's method. Two additional constant capacitors were added between the gate and the source/drain contacts to account for the overlap capacitances.

The different reliability evaluation algorithms that were presented in Section II were used in conjunction with a simple a-Si:H amplifier circuit from the literature [14] (see Fig. 8), to demonstrate the impact of the various algorithms on the results of the reliability simulation. The amplifier input is connected to a photodiode which provides a 4 nA pulse for 15 msec, with a period of 20 msec. The reliability simulation spanned an age of 20×10^7 seconds.[7] Fig. 9 shows the value of the threshold voltage versus circuit age as simulated using three simulation algorithms: the multi-step iterative, the multi-step non-iterative and the single-step non-iterative algorithms. To our knowledge, this work is the first to compare results of these three published reliability simulation algorithms. Fig. 10(a) and Fig. 10(b) show the transient voltage waveform at the output node of the amplifier at age equal 0.0 and 20×10^7 seconds, respectively, as computed using the multi-step, iterative algorithm. During this age period, a change of 3.8 V in the threshold voltage causes an increase of 3 V in the output voltage of the amplifier for the same input. Further simulation results were published elsewhere [15].

VI. Discussion

Having described the process of incorporating reliability simulation algorithms in an Analog-HDL

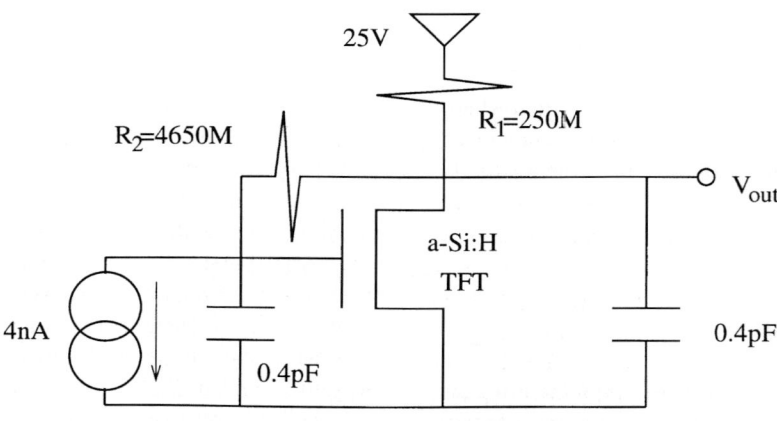

Fig. 8. An amorphous silicon amplifier.

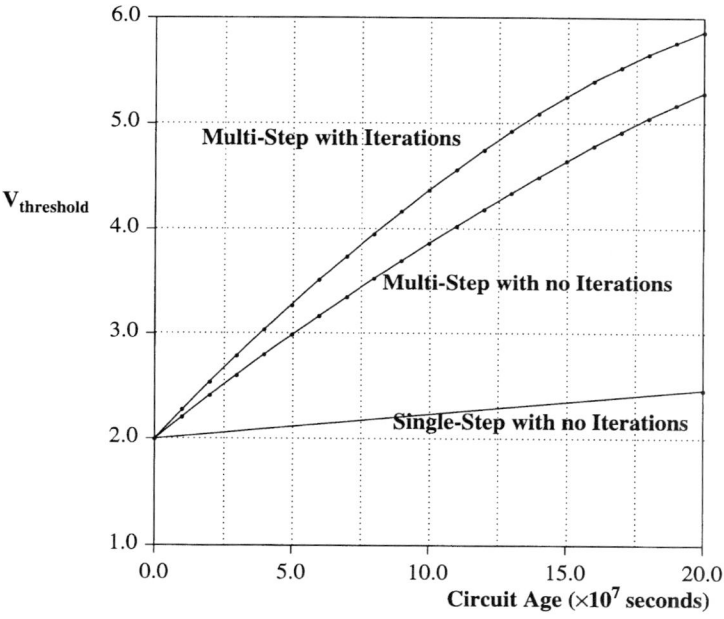

Fig. 9. Degradation in the threshold voltage for three different simulation algorithms.

environment using our methodology for implementing user-extended simulation algorithms, we now discuss other simulation algorithms that can be easily implemented into Analog-HDL environments (like VHDL-AMS).

The simulation of the effect of component failure on the performance of a system can be realized by adding extra models that describe the different failure mechanisms to the **component module** portion of the **event-driven electrical module**. In other words, each **component module** will contain one model that describes the proper operation of the component plus a number of other models that describe the different failure mechanisms which are possible for this component. Naturally, only one out of all these models is active at any given time. The choice of which model is active at any given time can be made by the **event-driven interface** under the direction of the user-extended algorithm. At the start of the simulation all **event-driven interfaces** ensure that the component model which represents the proper operation of the component is being used. Once the simulation is underway, the **event-driven control-modules** can instruct the **event-driven interfaces** of one or more components to switch from the proper operation model to a given failure mechanism model. This process can be repeated several times using consecutive time_slices to ensure adequate failure coverage.

Another simulation algorithm that can be implemented using our approach is the evaluation of the sensitivity of transient simulation results to the parameters of one or more components. The **event-driven control modules** can be designed to cause the **event-driven interface** to successively increment or decrement the parameters of one or more component within consecutive time_slices.

VII. Conclusions

We have presented an innovative simulation and modeling methodology that allows the implementation of user-extended simulation algorithms within generic Analog-HDL environments (like VHDL-AMS). The different algorithms are described in terms of a configuration which allows the implementation of user-extended simulation algorithms within a generic Analog-HDL environment using event-driven control modules, signals and interfaces. This configuration can

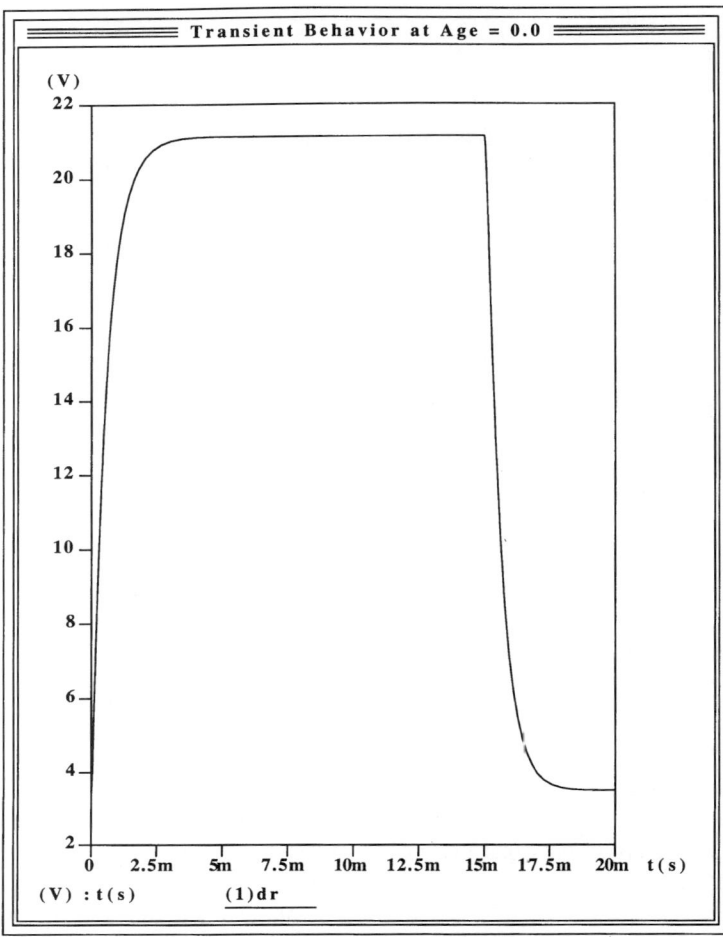

Fig. 10(a). Transient output voltage signal of the amplifier at age = 0.0 seconds.

be implemented within any generic Analog-HDL environment. Using three circuit reliability simulation algorithms as an example, we described the construction and operation of the event-driven control modules, signals and interfaces including IEEE 1076.1 PseudoCode listings of the important routines. We realized these reliability simulation algorithms within an existing Analog-HDL environment; this included the implementation of two concurrent versions of time within the same simulation. We presented the results of applying the three algorithms to amorphous silicon thin-film transistor circuits. Finally, we presented a discussion of other simulation algorithms that can be realized using our approach.

Acknowledgments

The authors gratefully acknowledge the Analogy University Program for the use of the Saber simulator in this work. The authors also thank C. McGuire for proof-reading the manuscript.

Notes

S. M. GadelRab was with the Department of Electrical and Computer Engineering, University of Waterloo, Waterloo, Ontario, Canada. He is now with Nortel Semiconductors, Ottawa, Ontario, Canada.

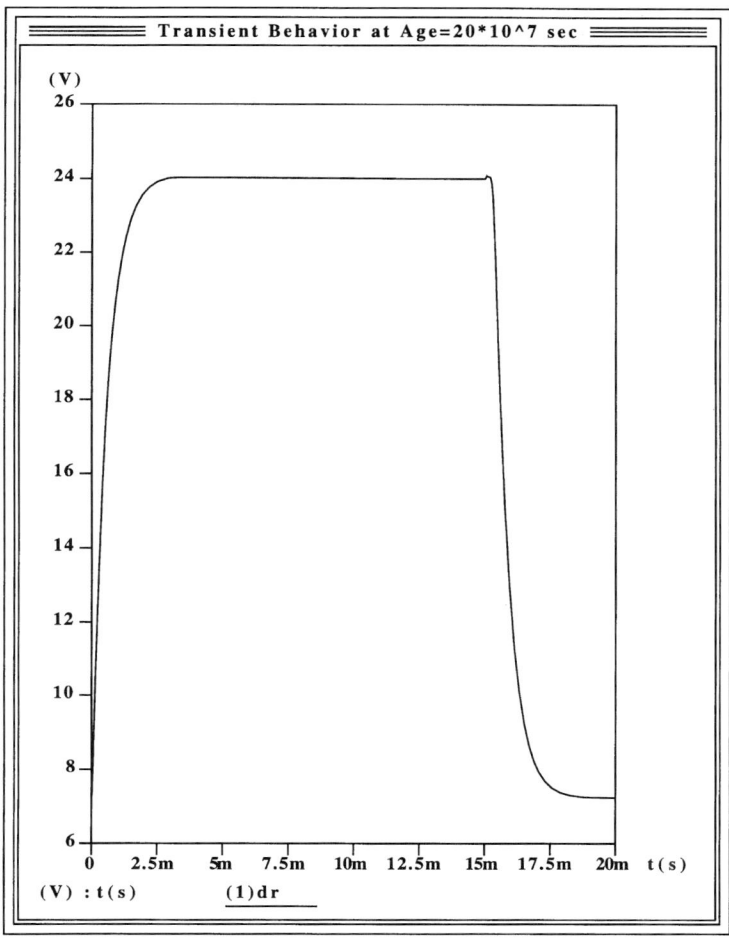

Fig. 10(b). Transient output voltage signal of the amplifier at age $= 20 \times 10^7$ seconds

J. A. Barby is with the Department of Electrical and Computer Engineering, University of Waterloo, Waterloo, Ontario, Canada. This work was supported in part by research funds arising from the Natural Science and Engineering Research Council (NSERC) of Canada (Grant OGP 000 3984) and the Information Technology Research Center (ITRC) of Ontario.

1. For example VHDL-AMS, ABCDL, HDL-A, MAST, SpectreHDL, Verilog-A, IsSpice4, etc.
2. A commercial circuit simulator from MetaSoftware, Campbell, CA.
3. In the iSMILE implementation, the addition of each degrading device parameter adds two extra circuit nodes per device instance thus causing a large increase in the system matrix.
4. The exact form of the degradation function is device dependent. An example of a degradation function is described in Section V.
5. Depending on how the BREAK command is implemented in VHDL-AMS, the extra integration time-steps may not be required.
6. The PseudoCode example uses a fixed-step approach to improve the readability.
7. This value of age is long enough to allow a clear distinction between the results obtained from different reliability evaluation algorithms.

References

1. H. A. Mantooth and M. Fiegenbaum. "Modeling with an Analog Hardware Description Language." Kluwer Academic Press, 1995.
2. R. Shi, E. Christen, P. Liebmann, S. Krolikoski and W. Zhou. "VHDL-A: Analog Extension to VHDL." *IEEE Int. ASIC Conf.* pp. 160–165, 1994.
3. R. A. Saleh, D. L. Rhodes, E. Christen, and B. A. A. Antao. "Analog Hardware Description Languages." *Proc. Custom Integrated Circuits Conf.* pp. 349–356, 1994.

4. W. Hsu, B. J. Sheu, and S. M. Gowda. "Design of Reliable VLSI Circuits Using Simulation Techniques." *IEEE J. Solid-State Circuits* 26, pp. 452–457, 1991.
5. P. M. Lee et al. "Circuit Aging Simulator (CAS)." *IEDM 1988* pp. 134–137, 1988.
6. K. N. Quader et al. "A Bidirectional NMOSFET Current Reduction Model for Simulation of Hot-Carrier-Induced Circuit Degradation." *IEEE Trans. Electron Devices* 40, pp. 2245–2254, 1993.
7. K. R. Mistry et al. "Circuit Design Guidelines for n-Channel MOSFET Hot Carrier Robustness." *IEEE Trans. Electron Devices* 40, pp. 1284–1295, 1993.
8. Y. Leblebici and S. Kang. "Modeling and Simulation of Hot-Carrier-Induced Device Degradation in MOS Circuits." *IEEE J. Solid-State Circuits* pp. 585–595, 1993.
9. A. T. Yang and S. Kang. "iSMILE: A Novel Circuit Simulation Program with Emphasis on New Device Model Development." *Proc. 26th Design Automat. Conf.* pp. 630–633, 1989.
10. J. Bhasker. "A VHDL Primer." Revised Edition, Prentice Hall, Englewood Cliffs, NJ, 1995.
11. J. Vlach, A. Opal and J. Wojciechowski. "Analysis of Nonlinear Networks with Inconsistent Initial Conditions." *IEEE Trans. Circuits and Systems* pp. 195–200, 1995.
12. H. A. Mantooth and M. Vlach. "Beyond SPICE with Saber and MAST." *Proc. IEEE Int. Symp. on Circuits and System* pp. 77–80, 1992.
13. M. J. Powell, C. van Berkel, and J. R. Hughes. "Time and Temperature Dependence of Instability Mechanisms in Amorphous Silicon Thin-Film Transistors." *Appl. Phys. Lett.* 54, pp. 1323–1325, 1989.
14. M. Hack, J. G. Shaw, and M. Shur. "Development of SPICE Models for Amorphous-Silicon Thin-Film Transistors." *Mat. Res. Soc. Symp. Proc.* 149, pp. 233–238, 1989.
15. S. M. GadelRab, J. A. Barby, and S. G. Chamberlain. "An Architecture for Integrated Reliability Simulators Using Analog Hardware Description Languages." *Proc. IEEE Int. Symp. Circuits and System* pp. 897–900, 1995.

Serag M. GadelRab received the B.Sc. degree in Electronics from Cairo University, Egypt, in 1990, and the M.A.Sc. and Ph.D. degrees in Electrical Engineering from the University of Waterloo, Waterloo, Ontario, Canada, in 1993 and 1996, respectively.

While at the University of Waterloo, his graduate research work involved the design, fabrication and modeling of amorphous silicon field-effect devices, as well as, the use of Analog Hardware Description Languages for predicting the reliability of amorphous silicon based circuits. He is currently with Nortel Semiconductors, Ottawa, Ontario, Canada, were he is working on behavioral modeling and verification methodologies for telecommunications systems.

James A. Barby received the B.Tech. degree in Electrical Engineering from Ryerson Polytechnical, Toronto, Ontario, Canada, in 1979, and the M.A.Sc. and Ph.D. degrees in Electrical Engineering from the University of Waterloo, Waterloo, Ontario, Canada, in 1981 and 1986, respectively. He joined the University of Waterloo's Electrical and Computer Engineering department in 1986. He has spent one year leaves with Analogy in Beaverton, Oregon (during 1990) and with Bell Labs in Allentown, Pennsylvania (Sep'95–Aug'96). He has worked with the IEEE 1076.1 language committee, validation committee, and working group. He has taught AHDL courses in Japan, at Dac'95, and ASIC'95. His research interests include simulation and modeling of circuits and systems including Analog Hardware Description Languages (AHDLs).

Analog Behavior Modeling and Processing Using AnaVHDL

LUN YE AND HAROLD W. CARTER[1]
University of Cincinnati, ECECS Department, P.O. Box 120030, Cincinnati, Ohio 45221-0030
email: hal.carter@uc.edu, lunye@thor.ece.uc.edu

Received October 24, 1995; Accepted November 26, 1996

Abstract. This article describes briefly the AnaVHDL language, which introduces minimal extensions to VHDL93 to construct a useful and robust language to model mixed-signal systems at multiple levels of abstraction from different equation level to abstract system level and across multiple technologies employing both discrete-event and continuous time behavior. The classification of equations to describe continuous domain system behavior is then presented. Based on the classification, a Compacted Nodal Analysis (CNA) method is used to construct the Differential/Algebraic Equation (DAE) set for a system design.

Key Words: mixed-signal modeling, hardware description language, compacted nodal analysis, modified nodal analysis

1. Introduction

VLSI technology has developed to the point that highly complex electronic systems are now integrated on a single chip. Moreover, analog blocks may reside on the same chip as digital components. Modeling and simulation of these mixed signal systems in a unified manner has thus become necessary.

Modeling and simulation technologies for purely digital systems range from device modeling to system behavioral modeling. While the benefits of behavior modeling for digital system have been major, modeling techniques for analog circuits have remained largely at the device level. But emerging mixed-signal modeling languages and simulators are beginning to directly handle mixed complexity circuits described at all levels, and with expected efficiency and accuracy exceeding present-day simulators.

The ever growing demand for a highly efficient and accurate simulation system for mixed analog/digital system behavior modeling has resulted in several related efforts to create an analog extension for VHDL, such as the AnaVHDL-I system [1], the ongoing IEEE 1076.1 standard (VHDL-A) [6] and MIMIC Hardware Description Language (MHDL) [8] (U.S. Army, 1991). AnaVHDL-I was an approach to mixed signal system simulation where the analog analysis routines and language processing functions were directly written in VHDL-87. AnaVHDL-I accepts a circuit specification in SPICE 2G6 format and performs DC and transient analysis with a precision similar to SPICE 2G6. The IEEE 1076.1 proposal is to add more language constructs to model the analog part of a mixed mode system. The MHDL is a functional language for modeling behavior of microwave systems.

AnaVHDL [11] uses minimal language modifications to VHDL-93 to support analog system modeling. VHDL constructs are fully capable of handling signals in the continuous time domain (with some syntax and semantic modifications). Interestingly, VHDL has many characteristics which are equally applicable to both discrete and continuous mode modeling. Thus, very little needs to be added to VHDL to support mixed-signal modeling.

Furthermore, a language like AnaVHDL is very useful for exploring new concepts in mixed-signal modeling, elaboration processing and mixed continuous/discrete simulation. Interestingly, AnaVHDL is semantically compatible with the emerging VHDL-AMS mixed-signal language standardization effort. Thus, concepts arising from the AnaVHDL project can aid the development of simulation systems for VHDL-AMS.

In this paper, we first give a brief introduction to

the AnaVHDL language. We then classify the equations used to model continuous domain behavior to ease the construction of the system Differential and Algebraic Equation (DAE) set for a system been modeled in AnaVHDL. A new memory-efficient analysis method, called Compacted Nodal Analysis (CNA) method, is then discussed and the application of CNA method to construct system DAE set is demonstrated. In the final section we present some issues that need to be addressed in future research and some directions that may be investigated for the design of a complete mixed-signal system modeling and simulation environment.

2. The AnaVHDL Language

To model continuous domain behavior, AnaVHDL adds several new elements to VHDL-93. The two primary additions are *analog* signals and *equation* statements.

2.1. Analog Signals

In AnaVHDL a new kind of signal, called the analog signal, is added to VHDL-93 to represent a continuous domain entity. Analog signals are candidates for unknowns in the system DAE set. Interface signal can also be of analog kind. While VHDL port signals can have a *mode* (i.e. a structure that specifies the direction of information flow through a port or parameter), analog port signals have no *mode* since they are intrinsically bidirectional.

The declaration syntax for an analog signal is:

```
signal_decl ::= SIGNAL id_list :
subtype_indic [kind] conversion_-
spec [:= expression];
```

where

```
kind      ::= REGISTER | BUS | ANALOG
tolerance_spec
```

tolerance_spec is used to specify the error tolerance for an analog signal. The conversion_spec is used to convert analog quantities to digital and vice versa.

Analog signals can be used to model analog quantities such as nodes in electronic circuits or torque in an electric machine model. However, analog signals are not the same as nodes in electronic circuits. For example, for electronic circuits, KCL should be enforced on each node. For analog signals, KCL needs only to be enforced for those analog signals that are nodes in the circuit. Analog signals can be used to model signal-flow behavior of some continuous system, such as the angular displacement of a damped free-motion pendulum. In this case, the analog signal representing the angular displacement of the damped free-motion pendulum is very different from a node in some electronic circuit, and it does not have the semantics of an analog signal that represents a node in a circuit.

AnaVHDL introduces several additional attributes over VHDL-93. The most significant ones are differentiation and integration of analog signals which takes the basic forms of As'DDT and As'INT, respectively, where As is an analog signal.

In the modeling of analog systems, there are two quantities that are of special interest. They are called *effort* and *flow*. "Effort" is defined to be the difference between some value associated with two nodes in a circuit. For example, voltage in an electrical circuit is an effort quantity since it is the difference between the potential at two nodes in the circuit. "Through" is defined as a quantity which flows from one node to another, for example, current in an electrical circuit. Other examples of effort and through quantities are similarly defined. In thermal analysis, the effort and flow are the temperature difference and the heat flux, respectively. In mechanical systems, effort and flow are torque and angular velocity, respectively. In electronic circuits. If the effort and flow are used with respect to a two-port component (represented by two analog signals in AnaVHDL), they are usually called "across" and "through", respectively.

Two special forms of analog signal usages are defined to describe the "across" and "through" properties between (two) analog signals. For example, in electronics, if Asa and Asb are two analog signals of a record type that has the real typed fields v and i, then the voltage [Asa-Asb].v denotes the "across" value between signals Asa and Asb, and the current [Asa:Asb].i denotes the "through" value between signals Asa and Asb. If Asa and Asb are the analog signals representing the two terminals of a capacitor, [Asa-Asb].v is the voltage across the capacitor, [Asa:Asb].i is the current through the capacitor.

The behavior of the capacitor is then described in AnaVHDL as

$$([Asa:Asb].i == C*[Asa-Asb].v'DDT);$$

For convenience, we explicitly denote a reference point for all the "across" properties in a circuit by GND. The "across" property of an analog signal is by default relative to the reference point in a system. For this case we only need refer to one signal to specify a particular "across". For example, [Asa-GND].v could be written as [Asa].v, or simply as Asa.v. The "through" property, however, cannot be represented as a single analog signal. One must specify the "through" property relative to two analog signals, flowing from the first analog signal to the second analog signal.

AnaVHDL uses labels to distinguish parallel "through" properties between one pair of analog signals. In electronics, parallel "through" properties with different labels between one pair of analog signals represent parallel electronic devices connected between the two analog signals, with each "through" denotes a two-terminal electronic device.

An analog signal does not have to have the "through" property or the "across" property. It can simply be an entity representing some information. The name of an analog signal is by default an "across" property if the signal is a scalar signal.

We say a design has explicit flow constraints when there is at least one pair of analog signals that have an explicit "through" property defined. For example, in the AnaVHDL description

$$([Asa:Asb].i == C*[Asa-Asb].v'DDT)$$

for a linear capacitor, [Asa : Asb].i is an explicit flow constraint that defines the "through" quantity between the two analog signals Asa and Asb.

We assume that the general nodal analysis method is used to evaluate mixed-signal system. That is, the simulator will compute the "across" property for all analog signals. In the general nodal analysis method, the circuit must observe Kirchhoff's Current Law (KCL). If explicit flow constraints are present for a design, the elaborator needs to enforce general KCL for the entire design. If no explicit flow constraints are present for a design, that is, there are no "through" signals in the model, the collection of all explicit equations forms the system DAE set.

2.2. Equation Statements

The equation statement is another construct AnaVHDL uses to model mixed-signal systems. In AnaVHDL, algorithmic statements and simultaneous differential/algebraic equations are used to model behavior. Besides assigning some value to an analog signal, the value of unknowns involved in simultaneous equations can also be determined by solving the system DAE set.

The following syntax is used to specify an equation statement:

```
equ_stat ::= ( expr == expr ) ;
```

An AnaVHDL equation is called a user-specified equation, or *explicit* equation. We consider an equation to be a mathematical model, whereas an equation statement in AnaVHDL is the mathematical model expressed in AnaVHDL syntax. In AnaVHDL, a design model can be constructed by using the procedural statements found in VHDL, or by using user-specified equations that describe the user designed system's behavior in a denotational manner. For an equation statement to be meaningful in AnaVHDL, there must be at least one analog signal in the equation statement.

Equation statements may appear as VHDL concurrent statements in an architecture body or in a process body where sequential process statements may appear. Equation statements have an independent execution semantics, that is, all equation statements are solved during simulation using continuous simulation methods similar to those in analog circuit simulators. The usage of identifiers (such as constants, signals, variables) in an equation statement must observe the scope and visibility rules of VHDL-93.

We now give two simple examples described in AnaVHDL. The first example, shown in Fig. 1(a), defines a serial RLC circuit with a sinusoidal voltage source. The second example, shown in Fig. 1(b), describes the displacement of a block sliding over a rough surface. More complex mixed-signal system models may be found in [11].

In order to describe a complete design, *implicit* equations may need to be constructed. Implicit equations are constructed by the simulator to make the DAE set complete such the solution of the DAE conserves the item in interest (e.g. charge for electronic circuits, general KCL).

As explained above, the DAE set of a design

```
ENTITY RLC IS
    GENERIC (R, L, C : REAL; Vs: REAL);
    PORT    (c1, c2 : electrical ANALOG);
END RLC;

ARCHITECTURE behavior OF RLC IS

    CONSTANT omega : REAL := 10.0;
    SIGNAL Vsin : electrical ANALOG;

BEGIN
  PROCESS
    BEGIN
        ( [Vsin - GND].v == Vs * SIN(omega * NOW) );
        ( [Vsin : c1].i  == [Vsin - c1].v / R );
        ( [c1 : c2].i    == C * [c1 - c2].v'DDT );
        ( [c2 - GND].v   == L * [c2 : GND].i'DDT );
    END PROCESS;
END behavior;
```

Fig. 1(a). A serial RLC circuit in AnaVHDL.

consists of explicit equations and implicit equations. The simultaneous equation set must be sufficient to specify the complete description of the system (to determine the values of all unknowns).

If a mixed-signal system modeled in AnaVHDL has no explicit flow constraints, only the explicit equations then forms the system DAE set, no implicit equations are created or used. Solution of this

```
ENTITY slipper IS
   GENERIC (mass:REAL:= 1.0;
slip:REAL:= 0.001; stick:REAL:= 0.01);
   PORT (external_force, position:REAL
   ANALOG);

END slipper;

ARCHITECTURE brick OF slipper IS
   SIGNAL effective_force, accelera-
tion, velocity : REAL ANALOG := 0.0;
SIGNAL stopped : BOOLEAN;

BEGIN

   ASSERT mass /= 0.0 REPORT mass'in-
stace_name & ''must not be zero.'';

   PROCESS
     BEGIN
       state1:
         stopped <= TRUE;
         WAIT ON external_force'CROSS
(stick), external_force'CROSS(-stick);
-- 'CROSS is a VHDL-A notion.
       state2:
         stcpped <= FALSE;
         WAIT ON velocity'CROSS(0.0)
```

```
          BEGIN
            IF stopped THEN
                    var1 := 0.0;
                    var2 := 0.0;
            ELSIF velocity > 0.0 THEN
                    var1 :=  1.0;
                    var2 := -1.0;
            ELSE
                    var1 := 1.0;
                    var2 := 1.0;
            END IF;
            ( position'DDT == velocity );
            ( velocity'DDT == acceleration );
            ( acceleration == effective_force / mass );
            ( effective_force == var1 * external_force + var2 * slip );
          END PROCESS;
      END brick;
```

Fig. 1(b). A slipping block model in AnaVHDL.

```
        UNTIL ABS(external_force) <
stick;
    END PROCESS;

    PROCESS (stopped)

    VARIABLE var1, var2 : REAL := 0.0;
```

DAE set is then the behavior of the continuous domain subsystem of the design. Since our interest is on systems with explicit flow constraints, we will restrict our discussion to those systems for the rest of this paper.

3. The Analog Model Processing Modules

Given a mixed-signal description of a circuit or any other system in AnaVHDL, the procedures in Fig. 2 are followed to process the analog portion of the circuit or system model. In this paper we focus our attention to the *equation classifying, canonical form translation* and *elaboration* procedures.

The equation classification module takes explicit equations in AnaVHDL text and matches them with the six different equation types described in the following section. The equations are processed first by the *symbolic permutation/matching* module such that names like `[a-b].v` and `[b-a].v` are recognized as the same with proper sign treatment. The *canonical form translation* module will apply some reduction operations described in the following section to further simplify the equation forms. The *elaboration* module will then construct the system DAE set based on the information gathered by the previous modules. Once the abstract form of the system DAE set is obtained, the *code*

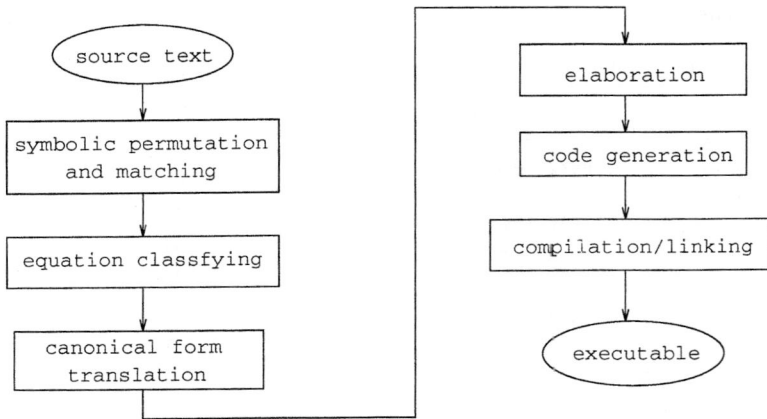

Fig. 2. An overview of the analog model processing modules.

generator will translate the abstract representation of the system DAE set into some intermediate language (like C++) or machine code.

4. Classification of Explicit Equations in AnaVHDL

The set of explicit equations that can be written in an AnaVHDL model can be classified into six different types, depending upon the presence and arrangement of the analog signals. The classification of explicit equations aids the elaborator to construct the DAE set for a design when it is necessary to construct implicit equations (to enforce general KCL).

4.1. Notations and Definitions

For ease of discussion we assume analog signals are of a record type that has two fields i and v.
1. Let $\mathscr{A} = \{a | a \text{ is an analog signal}\} \cup \{GND\}$
   That is, $\mathscr{A}$ is the collection of all analog signals.
2. Let $\mathscr{E} = \{[a - b].v | a, b \in \mathscr{A}\} \cup \{[a].v | a \in \mathscr{A}\} \cup \{a.v | a \in \mathscr{A}\}$
   $\mathscr{E}$ is the collection of "across" quantities. Since $[a-b].v$ and $[b-a].v$ differs only by a minus sign, we treat them as the same in the following discussion. Let $\mathscr{E}' = \{[a - b].v'DDT | a, b \in \mathscr{A}\}$. Similarly, define $\mathscr{E}^+ = \{[a-b].v'INT | a, b \in \mathscr{A}\}$.
3. Let $\mathscr{F} = \{[a : b].i | a, b \in \mathscr{A}\}$
   $\mathscr{F}$ is the collection of "through" quantities.

Again, $[a : b].i$ and $[b : a].i$ are treated as the same in the following discussion.
Let $\mathscr{F}' = \{[a : b].i'DDT | a, b \in \mathscr{A}\}$. Similarly, define $\mathscr{F}^+ = \{[a : b].i'INT | a, b \in \mathscr{A}\}$.
4. For any $\alpha = \{a, b\} \subseteq \mathscr{A}$, define
   $\alpha_e = [a - b].v$, $\alpha_f = [a : b].i$.
   $\alpha'_e = [a - b].v'DDT$. $\alpha'_f = [a - b].i'DDT$.
   $\alpha_e^+ = [a - b].v'INT$, $\alpha_f^+ = [a : b].i'INT$.
   Obviously, $\alpha_e \in \mathscr{E}$, $\alpha_f \in \mathscr{F}$. $\alpha'_e \in \mathscr{E}'$, $\alpha'_f \in \mathscr{F}'$. $\alpha_e^+ \in \mathscr{E}^+$, $\alpha_f^+ \in \mathscr{F}^+$.
5. For any $\alpha = \{a, b\} \subseteq \mathscr{A}$, $\beta = \{c, d\} \subseteq \mathscr{A}$, $\alpha_e \neq \beta_e \iff \{a, b\} \neq \{c, d\}$, $\alpha_f \neq \beta_f \iff \{a, b\} \neq \{c, d\}$.

For example, one can use $([a - b].v == K([c : d].i))$ to specify a CCVS in AnaVHDL notation. Using the definition above, that would be written as $\alpha_e = K(\beta_f)$, K being some function. We use the binary relation r to represent this mathematical notation and write it as $\alpha_e r \beta_f$. The constitutive equation of a linear capacitor would be written as $\alpha_e r \alpha_f$. Clearly, this relation is reflexive, symmetric (if K is invertible) and transitive.
The same notation holds for elements in $\mathscr{E}' \mathscr{E} e^+$, $\mathscr{F}'$, $\mathscr{F}^+$.
6. As an immediate application of the transitive property of relation r, we define four reduction operations on r. They are $f \to e$ reduction, $e \to f$ reduction, $e \to e$ reduction and $f \to f$ reduction. If $\alpha_e = K \times \beta_f$, and $\gamma_f = L \times \beta_f$, applying $f \to e$ reduction on β_f yields $\gamma_f = L/K \times \alpha_e$.
The other reductions are defined in a similar manner.

The same notation holds true for elements in $\mathscr{E}'$, $\mathscr{E}^+$, $\mathscr{F}'$, and $\mathscr{F}^+$.

Notice that in the above definitions, it is possible that not all "through" quantities in $\mathscr{F}$ are actually defined or meaningful in a design. For example, in electronic systems, the "through" quantity is meaningful only when it is defined for two adjacent nodes in the circuit. However, every element in $\mathscr{E}$ is either well defined by solving the DAE set and/or by direct assignment, or unconstrained (analog signals that are declared but not used).

4.2. Classifications of Explicit Equations

We categorize explicit equations into the following classes, according to the general form of each equation and the elements in $\mathscr{E}$, $\mathscr{E}'$, $\mathscr{E}^+$, $\mathscr{F}$, $\mathscr{F}'$, and $\mathscr{F}^+$ which are present in the equation. Classification of explicit equations will help the elaborator to construct the system DAE set for a particular AnaVHDL model, as we show below.

4.2.1. Class A Equations. Class A equations are essentially assignment statements in the form $\alpha_e = expression$, or $\alpha_f = expression$, where $\alpha_e \in \mathscr{E} \cup \mathscr{E}' \cup \mathscr{E}^+$, and $\alpha_f \in \mathscr{F} \cup \mathscr{F}' \cup \mathscr{F}^+$. To recognize this form of equations, reduction operations may be needed. These are degenerate equations. They may be used to model devices such as independent voltage or current sources. We will call an equation of the form $\alpha_e = expression$ an effort source and $\alpha_f = expression$ a flow source.

4.2.2. Class B Equations. Class B equations take the general form $\alpha_e r \alpha_f$, where $\alpha_e \in \mathscr{E} \cup \mathscr{E}' \cup \mathscr{E}^+$, and $\alpha_f \in \mathscr{F} \cup \mathscr{F}' \cup \mathscr{F}^+$. Examples of class B equations include the equation for a resistor, for a capacitor, or an inductor. However, class B equations are much broader than these passive components imply. For example, the equation $[a-b].v'DDT = [a:b].i'DDT$ is also included in class B. They in general specify the relationship of "across" and "through" properties for a pair of analog signals.

4.2.3. Class C Equations. Class C equations are equations in the form of $\alpha_e r \beta_f$, $\alpha_e \in \mathscr{E} \cup \mathscr{E}' \cup \mathscr{E}^+$, and $\beta_f \in \mathscr{F} \cup \mathscr{F}' \cup \mathscr{F}^+$. This class of equations may be used to model devices such as CCVS and VCCS. Class C equations describe a "controlled" relationship between analog signals. However, there must be a convention for identifying both the controlling factor and the controlled factor. An approach for determining these factors is underway.

4.2.4. Class D Equations. Equations in the form of $\alpha_f r \beta_f$ are class D equations. An example device described by this class of equations is the CCCS, i.e. flow source controlled by another flow. Class D equations includes, for example, equations of the form $[a:b].i'DDT = [c:b].i'INT$.

4.2.5. Class E Equations. Class E equations are equations in the form of $\alpha_e r \beta_e$. They describe devices that have a controlled "across"–"across" relationship, such as VCVS like devices, i.e. an effort source controlled by another effort. Since the "across"–"across" properties of the two pairs of analog signals are constrained by $\alpha_e r \beta_e$, we need only to compute the "across" of at most three analog signals by solving the system DAE set, and deduce the "across" for other analog signal(s), using the constraint $\alpha_e r \beta_e$.

4.2.6. Class F Equations. In class A, B, C, D and E equations we can only describe relatively simple system behavior. However, even the relatively simple equation forms cover all of the elementary circuit devices, such as resistors, capacitors, inductors, independent and controlled sources, etc.

AnaVHDL has a very liberal syntax for formatting equation statements. This makes it possible to use highly abstract mathematical equations to describe the system behavior at a very abstract level. However, with explicit flow constraints present in a design, it could be very difficult or even impossible to construct the system DAE set. In this case, either the designer needs to reformulate more complex equations into simpler forms so that the system DAE set can be constructed, or a more intelligent DAE construction algorithms must be invented.

If we can rewrite the explicit equations into a format such that an element in $\mathscr{F} \cup \mathscr{F}' \cup \mathscr{F}^+$ or in $\mathscr{E} \cup \mathscr{E}' \cup \mathscr{E}^+$ can be explicitly expressed as a function of entities that are in $\mathscr{F} \cup \mathscr{F}' \cup \mathscr{F}^+$ or in $\mathscr{E} \cup \mathscr{E}' \cup \mathscr{E}^+$, but do not include the element itself, the equations are of class F. The rewriting may just involve rearranging terms, applying the reduction operations defined earlier. However, there is no guarantee that this can always lead to a desired set of equations. No known solution to this dilemma yet exists.

In order to simply recognize controlling and controlled factors in a controlled source equation in classes C, D, E, we require the user to put the controlled factor on the left hand side of the symbol "= =", and the control function and controlling factor on the right hand side of the symbol "= =". Although this does impose a slight restriction on how equations can be written, this approach is intuitive.

Note that, even though we have used analog circuits to illustrate the concept presented here, they are not necessarily limited to circuit analysis. In fact, they can be similarly applied to other disciplines including mechanical, thermal, optical electromagnetic and others.

5. The Compacted Nodal Analysis (CNA) Method

Compare to the widely used Nodal Analysis (NA) method (see [9] for a good discussion on different circuit analysis methods), Modified Nodal Analysis [3] method and the lesser used Compacted Modified Nodal Analysis [4] method, CNA should result in smaller matrix sizes dependent primarily on the input description.

The CNA method is a modification of the NA method. The NA method cannot directly deal with circuits that have voltage sources or voltage controlled devices, except for VCCS. The CNA method handles all the cases except when the controlling factor of a current controlled device is the current of a voltage device. However, even in this case, the CNA method can handle it by simply adding the controlling factor as another unknown in the implicit equations.

5.1. System Graph Construction

The CNA method is used to construct the system DAE set. To explain the CNA method, we employ the aid of a circuit graph. Define an analog island to be a portion of the model such that there is no analog signal connecting it to the rest part of the model. Notice that if there are more than one analog islands in a model, there will be more than one graph, one for each analog island. It is interesting that since there is no direct analog connections between analog islands, analog islands can be processed easily in parallel if parallel simulator is used.

First, all analog signals in class A and B equations are collected. For each equation, an edge is drawn between the vertices representing the two analog signals in each equation. Vertices representing the same analog signals are merged into one vertex.

After the first step, all analog signals in class C, D and E equations are collected. For each equation, an edge is drawn between the vertices representing the two analog signals on the left hand side of the equation (controlled factor). As before, vertices representing the same analog signals are merged into one vertex.

In this paper, equation in class F are not permitted in AnaVHDL. Thus, we defer the method for creating the graph(s) for this type of equation until this restriction is relaxed. This restriction may make it inconvenient to model some high-level and complicated analog behavior for conserved systems, but it does not affect AnaVHDL's capability for the modeling of basic electric elements such as capacitors, inductors and analog behavior that can be modeled with class A through class E equations.[2]

We now use an example taken from [5] to illustrate the graph construction approach. Suppose we have the following explicit equations in AnaVHDL format where the shadowed box in Fig. 3 is a resistor (equation (9) in the following description). The corresponding graph is shown in Fig. 4.

```
( [a1:GND].i == a1.v * G1 );         -- (1)
( [a2: a1].i == [a2-a1].v * G2 );    -- (2)
( [a1:GND].i == I1 );                -- (3)
( [a1: a2].i == [a1-a2].v * G4 );    -- (4)
( [a2: a3].i == I3 );                -- (5)
( [a2:GND].i == [a2].v * G3 );       -- (6)
( [a3: a2].i == [a3-a2].v * G5 );    -- (7)
( [a3:GND].i == I2 );                -- (8)
( [a3:GND].i == [a3].v * G6 );       -- (9)
```

5.2. DAE Set Construction

When explicit flow constraints are present in a design, the elaborator must augment the system DAE set (created by the explicit equations) by creating implicit equations to enforce KCL. The specific process for

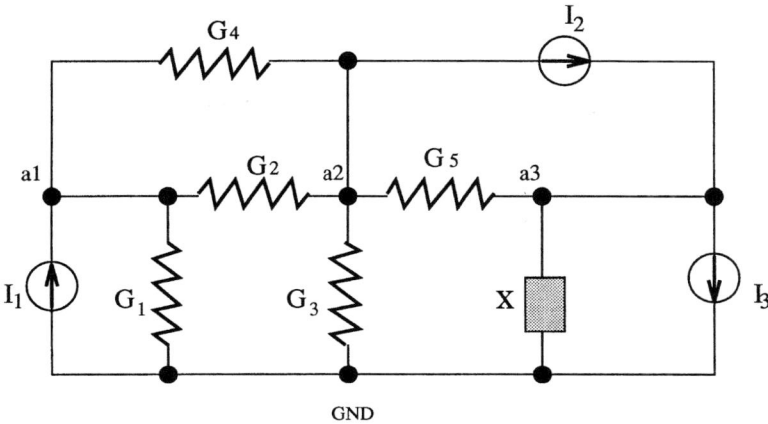

Fig. 3. Circuit example.

determining the implicit equations of the system DAE set is presented in this section.

An example of applying the CNA method is shown later in Section 5.3. For the CNA method, the following observations are made relative to electronic circuit analysis, but are also valid for any other disciplines where "through" and "across" elements are defined and a conservation basis is intrinsic to the system of components (for example, thermal system, mechanical system, etc).

1. At most one node voltage is solved for an independent or dependent voltage source. The voltage for the other node is evaluated in a post-processing phase according to the voltage source description since the two nodes' voltage differs by a simple and known expression. The matrix size is thus reduced by one over Nodal Analyses method.
2. The current of a voltage source is not considered as an unknown as long as it is not the controlling current of a current controlled device. When the current of a voltage source is not the controlling current of a current controlled device, we collapse the KCL equations for the two nodes of the voltage sources. This cancels the voltage source current since the currents flow into each of the collapsed nodes sum to zero. Thus, the system DAE has one fewer equation, and one fewer unknown.
3. The branch current for dependent current sources is not considered as unknown. The branch current

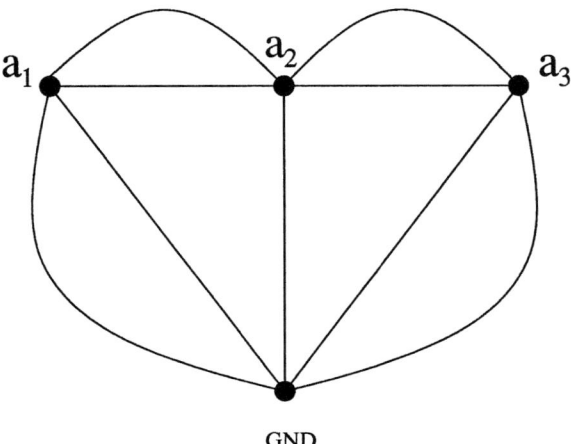

Fig. 4. The Graph for the Circuit in Fig. 3.

may be calculated from node voltages if the dependent current source is VCCS, or by KCL and/or reduction operations if the dependent current source is CCCS. This reduces unknowns more than the CMNA method which incorporates branch currents for dependent current sources as unknowns. Branch currents for dependent current sources are evaluated in a post-processing phase if they are output quantities.

5.3. Example

We now give an example to illustrate the DAE set construction process. For the circuit in Fig. 3, consider the component X to be a VCVS and replace equation (9) in the listing with $([a3].v = = K \times [a3 - a2].v)$. Also replace the independent current source I_1 with an independent voltage source V_1 with equation $(V_1.v = = w)$. Now we have only one unknown $[a2].v$ for the system since $[a1].v = w$, $[a3].v = K \times [a3 - a2].v$, which is $[a3].v = K/K - 1 \times [a2].v$. Once we know $[a2].v$ we know all the branch currents for resistors, and currents of the two voltage sources can be computed by enforcing KCL on $a1$ and $a3$.

As another example, if the component X is a CCCS, then equation (9) will be $([a3 : GND].i = = K \times [a2 : GND].i)$. Then the unknowns are $[a1].v$, $[a2].v$ and $[a3].v$. The implicit KCL equations for the three nodes are

$$[a1 - a2].v \times G_4 + [a1].v \times G_1 - I_1 - [a2 - a1].v \times G_2 = 0$$

$$[a2].v \times G_3 + [a2 - a1].v \times G_2 + I_3 - [a3 - a2].v \times G_5 - [a1 - a2].v \times G_4 = 0$$

$$[a3 - a2].v \times G_5 + K \times [a2].v \times G_3 - I_3 + I_2 = 0.$$

Notice that we used a $f \rightarrow e$ reduction on $[a2 : GND].i$ for the third equation. Now we have three equations, and during simulation we solve the three equations to get $a1.v$, $a2.v$, $a3.v$, followed by calculation of the branch currents (as needed).

In the process of implicit equation creation, sometimes explicit equations will be removed from the DAE set and used in the implicit equations. For example, the three equations shown above are the DAE set for the circuit in Fig. 3, and they are all implicit equations. The explicit equations (1) to (9) are all removed from the system DAE set, but the information they represent is incorporated into the system DAE set consisting of the three implicit equations.

6. Conclusions

So far AnaVHDL has shown significant promise in providing a wide-ranging and comprehensive capability to efficiently model and simulate mixed continuous/discrete system including mixed analog/digital electronic circuits. Key to this capability is the notion of classifying the types of equations permitted in AnaVHDL with a view towards formalizing the elaboration and code generation of these equations (see Fig. 2).

The structure of a component based system (system described hierarchically) is usually expressed explicitly. However, for a purely (largely) behavioral description, the system's structure is largely lacking and, thus, the simulation structure must be determined implicitly. Equation classifying aids the implicit determination of system structure necessary for simulation.

It appears that the Compacted Nodal Analysis method for constructing the system DAE set is capable of analyzing a wide range of devices than NA, usually with a smaller system DAE set (fewer equations).

Notes

1. This research sponsored by the Advanced Research Projects Agency (ARPA), contract J-FBI-93-116.
2. The limitation on class F equation is lifted and a detailed discussion can be found in [10].

References

1. S. Bangalore, W. Zhou, and H. Carter. AnaVHDL: A Mixed-signal Simulator Using VHDL. In *International Conference on simulation and Hardware Description Languages* pp. 43–47, January 1994.
2. David L. Barton. Issues in Mixed Mode Simulation Using VHDL and MHDL. In *International Conference on Simulation and Hardware Description Languages* pp. 37–42, January 1994.
3. C.-W. Ho et al. The Modified Nodal Approach to Network Analysis. In *IEEE Transactions on Circuits and Systems* CAS-22 of 6, pp. 504–509, June 1975.

4. GIELEN et al. A Symbolic Simulator for Analog Integrated Circuits. In *IEEE Journal of Solid-state Circuits* 24 of 6, pp. 1587–1597, December 1989.
5. Leon O. Chua et al. *Linear and Nonlinear Circuits.* McGraw-Hill, 1987.
6. VHDL Analog Sup-PAR Group. VHDL-A DoD and LES. anonymous ftp archive, 1995.
7. IEEE. *IEEE Standard VHDL Language Reference Manual*, IEEE Std 1076-1993 edition, 1993.
8. Intermetrics. *MHDL Language Reference Manual*, 1991.
9. Jiri Vlach and Kishore Singhal. *Computer Methods for Circuit Analysis and Design.* Van Nostrand Reinhold, New York, second edition, 1994.
10. Lun Ye. *A Language and Elaboration Method to Support Mixed-signal Modeling and Simulation.* Ph.D. Dissertation. ECECS department, University of Cincinnati, October 1996.
11. Lun Ye and Hal Carter. *AnaVHDL Language Reference Manual.* ECE/CS, University of Cincinnati, May 1995.

Analog and Mixed-Signal Extensions to VHDL

ALAIN VACHOUX

Integrated Systems Center, Dept. of Electrical Engineering, Swiss Federal Institute of Technology of Lausanne, CH-1015 Lausanne, Switzerland
E-mail: alain.vachoux@epfl.ch

Abstract. Hardware description languages (HDL) such as VHDL are today an essential technology to support most of the steps of digital hardware design, such as simulation, synthesis, testing, and formal proof. As the IEEE 1076 standard, VHDL is committed to evolve through five years re-standardization cycles whose objective is to make the necessary language changes or extensions in response to feedbacks from users and from tool suppliers. Requirements to support analog and mixed-signal systems have been issued during the initial phases of the second VHDL re-standardization cycle. Due to the complexity of the topic, a separate IEEE working group, referenced as 1076.1, was formally formed in 1993 with the charter to provide a language proposal based on VHDL 1076 that includes these new requirements. The language design phase is now complete and a solid language architecture is defined. A formal IEEE balloting process to approve the proposal as the new IEEE Standard 1076.1 has started in August 1997 and will be completed before the end of the year. This paper presents an overview of the 1076.1 language proposal that enhances VHDL to handle systems that exhibit continuous behavior over time and over amplitude. The way it is designed, VHDL 1076.1 will support the description and the simulation of both non-conservative and conservative continuous and mixed discrete/continuous systems.

Key Words: hardware description language, VHDL, mixed-signal

1. Introduction

Integrated circuit technologies have now reach a state where the design and the fabrication of more and more complex systems in a single chip is possible. This trend towards better performance and more complex chips has three main aspects.

First, the fabrication of complex digital systems uses submicron technologies in order to meet critical performance requirements such as area or speed. These systems are likely to exhibit analog behaviors as frequency rates are increasing and dimensions are decreasing. In particular, cell interconnections are behaving more and more as transmission lines. The primary digital functionality may hence become hampered by analog delays that cannot be detected unless specific accurate models are used during the design verification step [1].

Second, integrated circuits include more and more analog components such as A/D and D/A converters, phase locked loops, current mirrors, etc. Converters become necessary as integrated circuits are included in non-electrical systems such as mechatronic systems [2]. They become also necessary for pure electronic circuits for which some digital functionality is replaced by an equivalent, more powerful, analog one. Examples are: signal processing functions, filtering, high-frequency, low-noise, or neural applications [3].

Third, CAD tools are an essential means to ease the design process and to master the design complexity. CAD tools for digital design are now largely available and have reached a fairly high state of maturity. The emergence of VHDL (VHSIC Hardware Description Language), the IEEE Standard 1076 [4], as a unified medium to support a large part of the design process is an important contribution, as its actual use has extended over its initial goal (the description and the simulation of digital circuits and systems) to also address synthesis, formal verification, and testing [5,6]. On the contrary, CAD tools for analog design do not yet provide the same level of automation that it is possible now for digital design. This is mainly due to the large diversity of custom performance specifications analog circuits may require. On the verification side, simulation with accurate SPICE [7]-like electrical simulators is still widely used today because it allows the designer to stay closer to the final custom implementation. On the synthesis side, a lot of research is currently going on mainly in the academic

academic world in order to develop a coherent set of principles and algorithms [8].

The design of complex systems that include both digital and analog parts requires the use of appropriate descriptive means. Higher level descriptions using behavioral models become necessary as they allow for the simulation of the whole system in a shorter time than an equivalent transistor level description would do. Also, they are a typical input to automatic synthesis tools. A behavioral description of a block captures the necessary details (e.g. its terminal characteristics) without having to rely on any specific implementation. VHDL is already providing these capabilities in the digital domain. Some attempts have been made to exploit VHDL in the analog and mixed digital-analog domains, but with limited success as they had to cope with the underlying discrete time event-driven model that is intimately part of the language semantics [9][10].

This paper presents VHDL 1076.1, also informally called VHDL-AMS, a major effort to extend VHDL for the description and the simulation of analog and mixed-signal circuits and systems. The VHDL 1076.1 language has been designed within the 1076.1 Working Group of the Design Automation Standards Committee of the IEEE Computer Society. A language reference manual is currently reviewed through an IEEE balloting process and a first release of the IEEE Standard 1076.1 should be available by the end of 1997.

This paper is organized as follows. Section 2 gives a brief overview of VHDL. Section 3 presents an overview of the 1076.1 language. Section 4 gives some model examples. Finally, some concluding remarks are drawn.

2. VHDL Highlights

Before going into the VHDL 1076.1 language, it is worth to highlight the main aspects of VHDL to better understand the grounds on which the extended language has been designed.

VHDL is basically a discrete simulation language for discrete systems [11]. A VHDL model of a hardware system is viewed as set of concurrent processes that execute asynchronously according to events on signals. A *process* contains a set of statements that are executed in the given sequence once the process is triggered by an event on some sensitive signal. A *signal* is a VHDL object that represents a discrete time waveform whose amplitude may usually only take a value from a finite enumerated set of values. An event on a signal is deemed to occur when the signal value changes. A specific instruction within the process, the wait statement, allows to define the case(s) for which the process may trigger. Other VHDL objects include constants, variables and files that have the same meaning and use as in programming languages. Also, most VHDL *sequential statements* (i.e. those that may be only found within a process) resemble statements that exist in programming languages: e.g. conditional statements, loop statements, variable assignment statements, subprograms. Furthermore, VHDL process statements are themselves *concurrent statements* that don't need to be given in any specific order. Roughly speaking, VHDL sequential statements are used to describe algorithmic (or software) aspects of hardware systems, while VHDL concurrent statements are used to describe the concurrent behavior hardware systems naturally exhibit.

The way processes are triggered and executed, as well as the way signal objects get their values, are precisely and uniquely defined in the *canonical simulation cycle* that is part of the VHDL language definition. This guarantees portability: the same results are obtained from the same VHDL model whatever VHDL simulation tool is used. The canonical simulation cycle is event-driven. It uses a discrete time model where time is considered as an integral multiple of some base unit, the *resolution limit*. In VHDL, the minimum resolution limit is 1 fs. A larger resolution limit is possible to get a larger simulated time, at the price of a reduced time accuracy. Events may only occur, and therefore processes may only trigger, at some discrete time. Causality is ensured thanks to *delta delays*. Effects of signal assignments without any delay are propagated throughout the VHDL model at the same discrete time until signal values stabilize. Delta delays are iterations that ensure the same final stable point whatever order in which the processes are executed.

VHDL allows to encapsulate behavior into structure. The description of structure is based on the concept of *design entity*: a portion of hardware with well-defined inputs/outputs and performing a well-defined function. An *entity declaration* contains the interface aspects of the design entity, while one or several *architecture bodies* contain specific implementations of it. Interface aspects typically include the declaration of input and output connection points, called *ports*, and possibly some parameters, called *generic parameters*. Ports have a *mode* and a *type* that respectively define the direction and the kind of signals that are flowing through it. The compiled version of any design entity resides in a *design library*. The description of a hierarchical design uses *component declarations* and *component instances*. A *configuration declaration* binds the component instances to their actual models stored in some design library. The *configuration* mechanism, as well as the separate compilation of design entities, allows for the description and the simulation of several different implementations for the same part with a minimum recompilation effort.

Finally, VHDL also provides another design entity, the *package*, that allows for the encapsulation of common declarations and subprograms.

3. VHDL 1076.1 Language Overview

The VHDL 1076.1 language extends the existing VHDL 1076 language to satisfy a number of *design objectives* [12] that have been identified from requirements submitted during the second restandardization phase of VHDL. Design objectives have been prioritized according to their level of importance. A core set of essential objectives are met in the language proposal that is currently submitted to IEEE standardization. Another set of design objectives, that would either add too much complexity or conflict with the VHDL philosophy, have been postponed to a subsequent release of the language.

The 1076.1 language is targeted towards the description and the simulation of lumped analog, digital, and mixed-signal (digital-analog) hardware systems. Primary application areas are IC design, ASIC design, PCB applications, and electrical system design. However, the language is designed in such a way that non electrical systems (e.g. mechanical or thermal systems) are also supported to some extent.

3.1. VHDL 1076 Compatibility

The 1076.1 language is a strict superset of VHDL 1076. The consequences are twofold. First, any legal VHDL 1076 description is also a legal VHDL 1076.1 model and provides the same simulation results when simulated with either a VHDL 1076 or a VHDL 1076.1 simulator. This also means that the new mixed-signal simulation cycle in VHDL 1076.1 reduces to the VHDL 1076 canonical simulation cycle when no analog part is present in the simulated model. Second, the syntax and the semantics available in VHDL 1076 is reused as much as possible. As a result, some existing VHDL 1076 constructs are extended and new constructs are defined to support analog semantics. The approach in VHDL 1076.1 is to provide a clear-cut distinction between "digital" and "analog" statements, while keeping the general VHDL philosophy. As we'll see later, existing VHDL 1076 structural aspects are mainly extended, while new constructs are defined to deal with analog behavior. Moreover, the existing event-driven mechanisms in VHDL 1076 are reused to handle piecewise-defined and discontinuous analog behavior.

3.2. Behavior Aspects

Analog behavioral modeling is a key feature of VHDL 1076.1 as it allows the designer to develop his or her own models without having to rely on what a particular simulator is providing. It also allows a complete control on the abstraction level used for a particular model as well as on the characteristics of the system (e.g. nonlinearities) the model actually accounts for. VHDL 1076.1 provides a full support to the description of differential algebraic equations (DAEs) in the time domain of the form

$$F(\dot{x}(t), x(t), t) = 0 \quad (1)$$

where F is a vector of linear or nonlinear expressions, x is a vector of unknowns, $\dot{x}$ is a vector of first order derivatives of the unknowns w.r.t. time t. Unknowns are usually physical quantities that depend on the discipline of the system being modeled. They are voltage or current for electrical systems, velocity or force for mechanical systems, etc. The 1076.1 language provides a notation to express DAEs and it defines the equations that are implied by the text of a model. It does not, however, define how these equations have to be assembled to form the system (1), nor does it define how this system is actually solved. This choice is deliberate as many different equation formulation methods and simulation algorithms can be applied. The language assumes therefore the existence of an analog solver that is responsible for the computation of the unknowns in system (1). It also assumes that the analog solver computes a sequence of solutions, called analog solution points (ASPs), in some time interval. Since, in all generality, the values of the computed unknowns cannot solve the system (1) exactly, the 1076.1 language provides a way to control the numerical accuracy of the solution through the specification of tolerances.

3.2.1. Quantities. Quantities belong to a new class of objects (in addition to the existing constant, variable, signal, and file objects) that represent the unknowns in the system (1). Quantities are piecewise continuous waveforms function of time that may only take floating-point values. Quantities get their values from the analog solver at specific times, the so-called analog solution points or ASPs, that are determined by the analog solver. None of the other objects (i.e. variables and signals) get their values the same way. Quantities have to be declared before being referenced. They can be declared anywhere a signal declaration is allowed, except in a package. The following statement (1076.1 reserved keywords are typed boldface) declares two quantities, also called free quantities, q1 and q2 of type real:

quantity q1, q2: real;

The language also defines implicit quantities, i.e. quantities that don't need to be declared, that are required to writing DAEs.
- *Q'Dot* is an implicit quantity that is the first derivative of the (explicit) quantity Q with respect to time.
- *Q'Integ* is an implicit quantity that is the integral over time of quantity Q from time zero to the current time.

This notation can be "cascaded" to denote, for example, a second time derivative as *Q'Dot'Dot*, or any combination of the above attributes.

Other predefined implicit quantities will be introduced later.

3.2.2. Simultaneous Statements. Simultaneous statements are a new class of statements that are used to express differential and algebraic equations in the text of the model. Equations may be true input/output relationships or conservative branch equations.

In the first case, there is a directional computational flow and only one equation is allowed per output signal. A typical example is a signal-flow description of a filter. Compact notations to specify a *s* or *z* transforms are provided (3.2.5). In the second case, there is not any computational flow, but rather a set of equations that have to be solved simultaneously. The set of equations is typically built from the expression of KCL/KVL laws at connection points in which each branch equation contributes.

Simultaneous statements may appear at the same level as VHDL concurrent statements. The basic form is the simple simultaneous statement, which has the following syntax:

[label :]
lhs_expression == rhs_expression;

Expressions *lhs_expression* and *rhs_expression* may be any legal VHDL numerical expression. The statement is symmetrical as both expressions may appear on any side. Quantities, including implicit quantities of the forms Q'Dot and Q'Integ, may appear on both sides of the "==" mark. Expressions may include constants, literals, signals and, possibly user defined, function calls.

For example, the constitutive equation of a linear capacitor could be written in the following way:

ic == C * uc'dot;

where *ic* and *uc* are two quantities and *C* is a constant representing the capacitance value.

Other forms of simultaneous statements include the simultaneous if statement, the simultaneous case statement, and the simultaneous procedural statement.

The *simultaneous if statement* and the *simultaneous case statement* allow the description of piecewise defined behavior. For example, the behavior of a transconductance stage with current limiter could be described with a simultaneous if statement as:

if vin > vmax **use**
    iout == imax;
elsif vin < vmax **then**
    iout == -imax;
else
    iout == gm * vin;
end use;

where *vin* is the input voltage, *iout* is the output current, *vmax* is the maximum allowed input voltage, *imax* is the maximum allowed output current, and *gm* is the transconductance value. Both vin and iout are quantities, while *imax* and *gm* would be typically static parameters.

Such statements offer a way to define different sets of equations depending on some condition or the value of some expression. In the most general case, the switch may occur dynamically, that is during simulation, so care must be taken to handle possible discontinuities correctly. One way would be to ensure that there is not any discontinuity in the declared quantities. In the last example, one would add the following assert statement that checks the values of parameters *vmax*, *imax*, and *gm*, and reports an error if the characteristic is discontinuous:

```
assert abs(gm * vmax) = abs(imax)
  report ''Discontinuous Iout =
    f(Vin)''
  severity Error;
```

The case when a discontinuity is introduced is discussed later (3.2.4).

The *simultaneous procedural statement* is basically a way to write down equations in sequence. For example, consider the behavior of a weighted summer that is governed by the following equation:

$$\text{vout} = \sum_{i=1}^{m} \beta_i \text{ vinp} - \sum_{j=1}^{n} \gamma_j \text{ vinm}$$

where *vinp* (resp.vinm) is the positive (resp. the negative) voltage input, *b* (resp. g) is the vector of the *m* positive (resp. n negative) weights, and *vout* is the output voltage. Equation (2) may be then expressed as the following simultaneous procedural statement:

```
procedural
  variable bvs, gvs: real := 0.0;
  begin
  for i in beta'range loop
    bvs := bvs + beta(i) * vinp(i);
  end loop;
  for i in gamma'range loop
    gvs := gvs + gamma(i) * vinm(i);
  end loop;
    vout := bvs - gvs;
end procedural;
```

It should be noted that the same behavior could also be written with a single simple simultaneous statement, provided that the multiply operator "*" is overloaded to implement the dot product.

3.2.3. Tolerance Groups. The analog solver is responsible to compute the values of the quantities involved in simultaneous statements such that the relationships hold for all ASPs. Actually, the analog solver has to compute the values of quantities as to make the implied relations |lhs_expression - rhs_expression| as close as possible to zero.

To control this aspect, and to remain neutral with respect to the simulation algorithms, the language introduces the concept of *tolerance groups*, i.e. sets of quantities that have to meet the same accuracy requirements. VHDL 1076.1 allows to associate a string name, called a *tolerance code*, to tolerance groups. Tolerance codes are associated to quantity declarations as in the statements:

```
subtype voltage is real tolerance
  ''low_voltage'';
quantity qv: voltage;
-- indirect tolerance code association
quantity qc: real tolerance ''charge'';
-- direct tolerance code association
```

By default, all quantities belong to the tolerance group that is named with an empty string. Implicit quantities such as qv'Dot or qc'Dot belong to the same tolerance group as their "parent" quantity. Every simple simultaneous statement has a default tolerance group that is the one of the single quantity that appears on its left-hand or right-hand side, in that order. In the case the form of the simple simultaneous statement does not allow to infer the tolerance group, or in the case one would want to overload the default value, an explicit tolerance group may be specified in the statement as in:

```
ic == C * uc'dot tolerance
  ''other_tol'';
```

The interpretation of the tolerance code is left to the simulator, i.e. the simulator environment must provide a way to associate tolerance codes to actual tolerance values that depend on the simulation algorithms in use.

3.2.4. Mixed-Signal Interactions. Mixed-signal modeling and simulation involves the description and the simulation of a model that contain parts whose behaviors are either digital (or discrete, or event-driven) or analog (or continuous) [13]. VHDL 1076.1 allows the description of analog behavior (simultaneous statements) that is dependent on the values of digital signals, as well as the description of digital behavior (processes) that is dependent on, or sensitive to, the values of analog quantities. A typical example of the first digital-to-analog kind of interaction is the reference of a signal name in a simple simultaneous statement. A typical example of the second analog-to-digital interaction is a process that is triggered when a quantity crosses some threshold in the rising direction. Other typical mixed-signal interactions involve conversions between digital values (e.g. four state logic values) and analog values (e.g. voltage or current sources with internal impedances).

The 1076.1 language defines a new (digital) concurrent statement, the *break statement*, to deal with digital-to-analog kind of interactions that introduce discontinuities in the analog behavior. For example, the following portion of code describes the conversion of a two state logic value signal s to a floating-point quantity q:

```
if s = '0' use
    q == 0.0;  -- low input signal
else
    q == 5.0;  -- high input signal
end use;
break on s;  -- announce discontinuity
```

The break statement in this example is equivalent to a process that is triggered whenever the signal s has an event. The backside effect is to set an invisible flag (i.e. not accessible to the model) the analog solver will sense to detect that a discontinuity has occurred. The analog solver is this way told to compute the new state of the analog part of the model after the discontinuity. The break statement is mandatory if a signal is referenced in an expression in a simple simultaneous statement. Note that there may be cases where any event on the signal does not introduce any discontinuity. The break statement is also mandatory if a discontinuity is introduced by the selection of simultaneous statements, as in the previous example. VHDL 1076.1 considers the model to be erroneous if no break statement is specified when one of the two above cases happens.

The 1076.1 language however defines two new implicit quantities that alleviate the need to specify an explicit break statement when a digital-to-analog interaction has to occurs. Given a signal S, the following predefined attributes are introduced:

- The quantity $S'Ramp(Tr,Tf)$ follows the signal S, but ramps linearly over the rise time Tr or fall time Tf when S has an event to the new value of S. Either Tr or Tf may be zero, In this case, the induced discontinuity is handled internally in the definition of the attribute.

- The quantity $S'Slew(RSlope\_Max, FSlope\_Max)$ follows the signal S, but ramps linearly with a rising slope, resp. falling slope, that is limited to RSlope_Max, resp. FSlope_Max, when S has an event to the new value of S. Either RSlope_Max or FSlope_Max may be omitted to denote an infinite slope. In this case, the induced discontinuity is handled internally in the definition of the attribute.

To cope with analog-to-digital interactions, the 1076.1 language defines a new predefined attribute on quantities of the form $Q'Above(E)$, which is an implicit boolean signal that has the value FALSE when the value of quantity Q is below the threshold E (E may be any floating-point expression), and the value TRUE when the value of the quantity Q is above the threshold E. An event on this signal occurs when the sign of the expression Q–E changes. For example, the following portion of code describes the conversion of the value of an analog quantity q to the corresponding value of a two state logic value signal *s*:

```
s = '1' when q'Above(high_level/2.0)
else '0';
```

where high_level would be a floating-point value denoting the analog level that corresponds to a logical one. Conversions between waveform values have to be explicitly defined in the model (e.g. with specific interface components). Default conversions, e.g. the direct association of event-driven and continuous time waveforms, are not supported in the language as they would add overly complex semantics.

3.2.5. Frequency Domain Modeling. VHDL 1076.1 does not provide any support for general frequency domain modeling because the language scope is restricted to lumped systems. Small-signal AC and noise simulation are however supported, as described in 3.3.3. The language also provides a notation to describe abstract filters as Laplace and z-domain transfer functions. Given a quantity Q, the following predefined quantity attributes are introduced:

- The quantity *Q'Ltf(Num,Den)* defines a Laplace transfer function with quantity Q as input. Num, resp. Den, are the polynomial coefficients of the numerator, resp. the denominator, of the transfer function expressed in the Laplace variable s.
- The quantity *Q'Ztf(Num,Den,T,Init_Delay)* defines a z-domain transfer function with quantity Q as input. Num, resp. Den, are the polynomial coefficients of the numerator, resp. the denominator, of the transfer function expressed in the variable z^{-1}. T is the sampling period and the first sampling occurs after Init_Delay seconds.
- The quantity *Q'ZOH(T,Init_Delay)* defines a zero order sample-and-hold with a sampling period T and a first sampling after Init_Delay seconds.

3.3. Simulation Aspects

VHDL 1076.1 extends the initialization phase and the simulation cycle as defined in VHDL 1076 to support mixed-signal initialization, quiescent state computation, user-defined initial conditions, mixed-signal time domain simulation, and frequency domain simulation (small-signal AC and noise simulation).

3.3.1. Initialization and Initial Conditions. VHDL 1076.1 defines the quiescent state of a mixed-signal model as the state at which there is no pending event at time zero. This includes the computation of the DC operating point for the analog part of the model. The computation of the quiescent state of the model may be informally described with the following simplified pseudo-code:

```
Set time = 0.0
Set all signals and quantities to their
initial values
Execute all non postponed processes
until they suspend
Execute all postponed processes until
they suspend
Apply initial conditions (if any)
repeat {
 Resume analog solver
 Update signals
 Execute active non postponed processes
 if no more pending events {
  if time domain simulation {
   Compute time Tn of next event
   Execute active postponed processes
  }
  if frequency domain simulation {
   Perform small-signal AC or noise
    calculation
  }
 }
} until no more pending events
```

Initial values on signals or quantities are defined in their declaration (the default initial value on quantities is 0.0, which is a reasonable value in many cases). They are used by the algorithm that computes the DC operating point for its first iteration. Initial conditions are given as equations in the source code with a particular use of the break statement:

break q => expression;

This concurrent statement is equivalent to a non postponed process that executes only once at the beginning of the initialization phase and then suspends forever. Its execution defines the new equation q - expression == 0.0 that will replace the implicit default equation q'Dot == 0.0 for the computation of the quiescent state. This can only happen when q'Dot explicitly appears in the text of the model. When it is not the case, another syntax is provided:

break for q1 **use** q2 => expression;

For example, an initial voltage v0 on a capacitor may be specified in either following way, depending on how the constitutive equations are written:

```
i == C * v'dot;
break v => v0;
-- or:
q == C * v;
i == q'dot;
break for q use v => v0;
```

From the point where the quiescent state is reached, several kinds of simulations are possible: (possibly mixed-signal) time domain, small-signal frequency domain or noise simulation. A new predefined signal DOMAIN, which can have one of the values INITIALIZATION_DOMAIN, TIME_DOMAIN, or FREQUENCY_DOMAIN, is introduced for two purposes: to detect when the quiescent state is reached, and to allow to write different behaviors depending on the kind of simulation.

3.3.2. Time Domain Mixed-Signal Simulation.

VHDL 1076.1 extends the event-driven simulation mechanism of VHDL to support analog and mixed-signal simulation. VHDL 1076 assumes the existence of the so-called kernel process that is responsible to coordinate process activities during simulation and to update signal values accordingly. Time is considered as an integral multiple of some base unit (e.g. 1 fs) and time directly advances to the next event that is scheduled in the future.

The analog solver solves the system of equations (1) with a variable floating-point time step that is determined by the simulation algorithms in use. The 1076.1 language precisely defines the synchronization between the VHDL event-driven simulation cycle and the execution of the analog solver thanks to a new unified model of time that encompasses the requirements for both discrete and continuous simulation. The unified model of time is based on an ideal floating point system with time values observable in either the physical type Time (predefined in VHDL 1076) or a floating point type.

Conversion functions are defined to convert between floating-point and physical time values with the equivalence 1 second of physical time = 1.0 floating-point time. The mixed-mode simulation cycle in VHDL 1076.1 may be informally described with the following simplified pseudo-code:

```
repeat {
 Resume analog solver
   (Tn may be reset to Tn' < Tn as a
result of an A-to-D event)
 Set current time Tc = Tn
 Exit if Tn = Time'High or no more
active processes
 Update signals
 Resume active non postponed processes
 Compute next time Tn
 Determine which analysis domain will
follow
   e.g. time domain or frequency domain
 If Tn = Tc then continue (delta
delay)
 Resume active postponed processes
 Compute next time Tn
}
```

The analog solver executes and computes a sequence of ASPs at the beginning of the next simulation cycle between times Tc (the current time) and Tn (the next time when an event will occur). In the case some process is sensitive to an implicit signal of the form Q'Above(E) and an event occurs on that signal, the analog solver has to prematurely stop executing and let the remaining part of the simulation cycle to perform. This forces the process, and all other processes in subsequent delta delays, to execute at a time, called an *offset time*, that may not be representable exactly as a value of physical type Time. Note that any regular non-zero delay signal assignment still occurs at an exact physical time.

The new mixed-mode simulation cycle reduces to the existing 1076 simulation cycle if the model does not include any quantities, and that it reduces to the single execution of the analog solver if the model does not include any signals.

3.3.3. Frequency Domain Simulation.

VHDL 1076.1 supports frequency domain simulation in that it defines how a small-signal AC simulation and noise simulation are carried out. It formally defines the small-signal model of system (1) as the linear incremental model obtained from its Taylor expansion. The resulting linear system of equations is then solved considering the application of frequency domain sources, which are defined in the text of the 1076.1 model as a special kind of quantities called *source quantities*.

Source quantities come in two flavors depending on the kind of frequency domain analysis. The *spectral source quantity* defines a stimulus for the small-signal AC simulation, while the *noise source quantity* defines a stimulus for the noise simulation. Both source quantities are handled as through quantities as defined in 3.4.1. A special syntax in the quantity declaration is provided to define such sources. A small-signal AC source quantity declaration defines the magnitude and the phase of the source, while a noise source quantity defines the magnitude of the source only:

```
quantity q_ssac: real spectrum
   magnitude, phase;
quantity q_noise: real noise magnitude;
```

The expressions magnitude and phase may be any general VHDL expression. They may include quantities and calls to the predefined function FREQUENCY that returns the value of the current simulation frequency.

3.4. Structure Aspects

The description of analog and mixed-signal structures basically reuses the same structuring mechanisms as found in VHDL 1076, i.e.: design units, interface definitions in entity declarations, implementations in architecture bodies, component declarations and component instances, and design libraries. However, modeling the interconnection of analog components has a very specific meaning, so new kinds of connection points are introduced.

3.4.1. Conservative Systems. Modeling conservative systems requires that conservation laws must be satisfied at connection points. These are Kirchhoff's Current and Voltage Laws (KCL/KVL) for electrical circuits. Generalized versions of KCL and KVL may also be derived for non-electrical systems. A conservative system is usually described using a graph-based approach involving two kinds of physical quantities, namely *across quantities* and *through quantities*, that are characteristic of the energy domain, or the engineering discipline, the system belongs to. For example, across quantities represent effort like effects such as voltage, temperature, or pressure, and through quantities represent flow like effects such as current, heat flow rate, or fluid flow rate, for the following electrical, thermal, and fluid energy domain, respectively.

The constitutive equations of conservative systems are then expressed by relating the across and through quantities of one or several branches. Such quantities are therefore called *branch quantities* in VHDL 1076.1. A branch is defined between two terminals that belong to the same nature. For example, a resistor has a single branch (hence, two terminals), and its constitutive equation (Ohm's law) relates the voltage across (the across quantity) and the current through (the through quantity) the resistor terminals. Such explicit relationships are expressed using the new class of simultaneous statements (3.2.2). Note that the equations that describe the conservative aspects have not to be explicitly written in the text of the model. They are implicitly derived from the structure of the model (i.e. from terminal and branch quantity declarations).

3.4.2. Natures. A nature in VHDL 1076.1 defines the properties of a terminal as a pair of floating-point types. As an example, here is the definition of the nature electrical:

```
subtype voltage is real;
subtype current is real;
nature electrical is
  voltage across
  current through;
```

Similar definitions can be made for other disciplines, such as mechanical, thermal, acoustical, etc. Several predefined attributes are provided to extract specific nature properties. Let N be a nature name such as *electrical*. Then, *N'Across*, respectively *N'Through*, denotes the type associated to the across, respectively through, quantities that are related to the nature N. Another attribute, *N'Reference*, denotes the reference terminal of the nature N (e.g. the electrical ground for nature *electrical*).

3.4.3. Terminals. The *terminal* is the second new object introduced in VHDL 1076.1. The main difference from other objects is that a terminal does not bear any value by itself. Therefore, a terminal does not have a type; instead it is defined to belong to a specific nature. For example, the following declarations declare two terminals of nature *electrical*:

```
terminal t1, t2: electrical;
```

A terminal may be declared anywhere a signal declaration is allowed. In particular, a terminal can be an interface element in a port declaration list. In that case, it is called a *terminal port*. For instance, the following entity declaration defines a simple interface of a diode component:

```
entity diode is
  port (terminal anode,
    cathode: electrical);
end entity diode;
```

The association of terminal ports is used to construct nodes in hierarchical descriptions in a fashion paralleling the use of signal ports to construct nets in digital hierarchies.

The declaration of a terminal T creates two quantities that are named using predefined terminal attributes. *T'Reference* is an across branch quantity defined between T and the reference terminal of the nature of T. *T'Contribution* is a through branch quantity whose value is equal to the algebraic sum of all through branch quantities incident to T (with appropriate sign).

The above definitions equally apply to any *local terminal*, i.e. any terminal that is declared locally to an internal block such as an architecture body.

3.4.4. Branch Quantities. A *branch quantity* is defined between two terminals, its plus terminal and its minus terminal. A branch quantity declaration is a specific quantity declaration that may define both across and through branch quantities. For example, if we reuse the terminal declaration for terminals *t1* and *t2* in 3.4.3, the following branch quantity declaration declares one across branch quantity *v12* and one through branch quantity *i12* between plus terminal *t1* and minus terminal *t2*:

quantity v12 **across** i12 **through**
    t1 **to** t2;

There are several points to make here. First, the types of the quantities *v12* and *i12* are derived from the across and the through aspects of the nature of the terminal, i.e. *v12* is of subtype *voltage*, and *i12* is of subtype *current*. Second, the definition of the plus and minus terminals also implies that the across branch quantity *v12* actually holds the difference *t1'Reference–t2'Reference*, and that the through branch quantity *i12* actually flows from *t1* to *t2*. Third, it is the declaration of a through branch quantity that actually defines a topological branch in the underlying graph. Having the following legal declaration:

quantity v12 **across** t1 **to** t2;

merely declares a branch quantity *v12* whose value senses *t1'Reference–t2'Reference*. Consequently, declaring a list of across branch quantities between the same terminals merely defines aliases for the same quantity. Another consequence is that defining a list of through branch quantities between the same terminals does declare as many *parallel topological branches*. The following declaration:

quantity v12 **across** i12a, i12b **through**
t1 **to** t2;

declares two parallel topological branches between terminals *t1* and *t2*. In all cases, if terminal *t2* is omitted in the declaration, the reference terminal of the corresponding nature is implied.

3.4.5. Signal-Flow Modeling. The description of signal-flow systems does not require conservative connection points such as terminals. VHDL 1076.1 allows to declare interface quantities in the entity port list. Such quantities are called *quantity ports*. For example, the following example models of a signal-flow adder:

entity sfgadder **is**
 port (**quantity** in1, in2: **in** real;
quantity sum: **out** real);
end entity sfgadder;
architecture sfg **of** sfgadder **is**
begin
  out == in1 + in2;
end architecture sfg;

Quantity ports have a mode which is restricted to be *in* or *out*. The meaning of the mode on quantity ports is different than from the one that already exists for signal ports. It defines legal quantity port associations and it provides a support to perform solvability checks on the description.

3.5. Miscellaneous Aspects

The VHDL 1076.1 language is a rich and powerful language that provides many other facilities. Some of them are briefly outlined in this section.

3.5.1. Time Step Control. VHDL 1076.1 provides a way to control the time step the analog solver is taking when computing a sequence of ASPs. Given an ASP at time Ti, a call to the new predefined function LIMIT_NEXT_STEP(Tmax) during the computation of that ASP forces the next ASP to be computed at a time Ti+1 such that Ti < Ti+1 Tmax. A call to this function may be done in any user defined function called from any simultaneous statement.

3.5.2. Solvability Checks. The 1076.1 language defines rules any model must satisfy to ensure that the system of equations (1) is solvable. A necessary, but not sufficient, condition is that there must be as many equations as unknowns. In a VHDL 1076.1 model, unknowns are represented by quantities and equations are represented by explicit simultaneous statements in the text and implicit simultaneous statements from the topological interconnection of

components and the conservative semantics. The consequences of the introduction of such checks in the language are:
- The definition of mode on quantity ports.
- In a design entity, there must be as many explicit simultaneous statements as declared through branch quantities, free quantities, and interface quantities of mode **out**.
- It is not allow to declare a free or a through branch quantity in a package.

Note that the existing scoping rules in VHDL make the detection easier as it is possible to perform the solvability checks at the individual design entity level.

3.5.3. Statistical Modeling. Statistical modeling includes two aspects: support for Monte Carlo simulations and support for time domain noise modeling (or more generally, time series modeling). The investigations in this area have shown that both aspects can be supported by specific packages, using the language elements that already exist in VHDL or that have been introduced in VHDL 1076.1 for other purposes. No new syntax is therefore needed. Following the VHDL philosophy, packages to support statistical modeling are not included in the language definition, but are rather candidates for a possible separate standardization (in the same way the nine state logic value system package STD_LOGIC_1164 has been standardized separately from VHDL 1076).

3.5.4. Mathematical Functions. Mathematical functions are not included in the 1076.1 language as they are available in the standard VHDL 1076.2 packages called MATH_REAL and MATH_COMPLEX. As a convention, a logical library called IEEE containing these packages, among others, should be available in any VHDL environment.

4. 1076.1 Model Examples

Here are some examples of VHDL 1076.1 descriptions are given that highlight the aspects presented in section 3. Only examples related to electrical systems are given here, but it is also possible to describe any kind of continuous dynamic systems whose behavior is expressed as a set of DAEs. Reserved 1076.1 keywords are typed boldface.

4.1. Basic Aspects: Diode

The simple model of a diode device element given in Example 1 illustrates many basic aspects of the 1076.1 language.

```
library IEEE, Disciplines;
use Disciplines.electrical_system.all;
use IEEE.math_real.all;
entity diode is
 generic (ISS, N, VT, TT, CJ0, VJ, RS:
real);
 port (terminal p, m: electrical);
end entity diode;
architecture level0 of diode is
 quantity vd across id, ic through
         p to m;
 quantity qc: real;
begin
 id= =ISS*(exp((vd-RS*id)/
         (N*VT))- 1.0);
 qc= =TT*id - 2.0*CJ0*sqrt
         (VJ**2- VJ*vd);
 ic= =qc'dot;
end architecture level0;
```

Example 1. Simple diode model.

The library *Discipline* contains the package *electrical_system* that defines the nature *electrical* (3.4.2). The library IEEE contains the package *math_real* that defines real-valued mathematical functions that are used in the model. The entity declaration defines a list of generic parameters and two terminals *p* (the anode) and *m* (the cathode). Both terminals are of nature *electrical*. The architecture body declares two parallel branches between the terminals *p* and *m*, whose currents are *id* and *ic*, respectively, and whose common voltage difference is *vd*. It also declares the free quantity called *qc* that does not belong to any terminal. Three simple simultaneous statements that describe the constitutive equations of the device model are then given. The use of the free quantity *qc* is here mandatory as it is not possible to express the time derivative of an expression. An alternative formulation that would not need quantity *qc* would be to replace the last two simultaneous statements with the following single statement:

```
ic'integ  = = TT*id -
2.0*CJ0 * sqrt(VJ**2 - VJ*vd);
```

The following equations are implicitly derived from the conservative semantics of the terminal and the branch quantity declarations:

```
vd == p'reference - m'reference;
p'contribution == id + ic;
m'contribution == -(id + ic);
```

4.2. Mixed-Signal Models: Generic A/D and D/A Converters

The model of an A/D converter is a typical mixed-signal model that involves both quantity and signal objects. The model in Example 2 is generic as it may be used to convert an analog input into an unsigned bit vector whose length is defined by the environment in which the component is instantiated.

```
library Disciplines;
use Disciplines.electrical_system.all;
entity adc is
 generic (vmax: real := 5.0);
 -- max. voltage at input
 port (terminal ain: electrical;
 -- electrical input
   signal clock: in bit;
 -- conversion control
   signal dout : out bit_vector);
 -- bit vector output
end entity adc;
architecture sa of adc is
 quantity vin across ain;
 -- across quantity to ground
begin
 process
   variable thresh: real;
 -- threshold to compare
   variable vsam: real;
 -- sample value of analog input
 begin
   wait until clock = '1';
   assert vin < vmax and vin >= 0.0;
 -- check input range
   thresh := vmax;
   vsam := vin;
   for i in dout'reverse_range loop
 -- range MSB:LSB assumed
     thresh := thresh / 2.0;
     if vsam > thresh then
       dout(i) <= '1';
       vsam := vsam - thresh;
     else
       dout(i) <= '0';
     end if;
   end loop;
 end process;
end architecture sa;
```

Example 2. Generic A/D converter model.

The entity declaration defines the generic parameter *vmax* and three ports: an analog port ain and two digital ports clock and dout. The architecture body implements a simple successive approximation algorithm on the sampled analog input. The branch quantity *vin* senses the analog input voltage, then a digital process is triggered on the rising edge of the signal *clock* to compute the output bit vector *dout*. The conversion will fail if the input voltage is outside the range $0.0 \leqslant vin < vmax$. This condition is checked with the assert statement.

The model of a simple D/A converter is another typical mixed-signal model. The model in Example 3 is also a generic model in that it converts a digital input signal of any length into an analog value that is realized as an ideal voltage source.

```
library Disciplines;
use Disciplines.electrical_system.all;
entity dac is
 generic (vmax: real := 5.0);
 -- maximum voltage
 port (terminal aout: electrical;
 - analog output
   signal din: in bit_vector);
 -- bit vector input
end entity dac;
architecture simple of dac is
 function bit_to_real (b: bit_vector)
         return real is
   variable weighted_sum: integer := 0;
 begin
   for i in b'reverse_range loop
     weighted_sum := 2*weighted_sum +
                 bit'pos(b(i));
   end loop;
   return real(weighted_sum) /
              2.0**b'length;
 end function bit_to_real;
```

```
  quantity vout across
           iout through aout; -- branch
           quantities to ground
begin
  break on din;
  vout == vmax * bit_to_real(din);
end architecture simple;
```

Example 3. Generic D/A converter model.

The entity declaration is straightforward. The architecture body implements equation (3) with the user-defined function bit_to_real.

$$vout = \frac{vmax}{2^N} \sum_{i=1}^{N} din(i) \cdot 2^i$$

The branch quantity declaration now declares both an across quantity *vout* and a through quantity *iout*. The through branch quantity is mandatory to define a topological branch between the output and the ground and to realize an ideal voltage source: the value of the through quantity *iout* will be computed by the analog solver such as the value imposed on *vout* is delivered at the terminal *aout* whatever the actual load on that terminal is. Additionally, the use of the break statement is mandatory as a discontinuity is introduced whenever the digital input *din* has an event. The explicit break statement could be avoided with the use of the predefined signal attribute S'Ramp. Example 4 gives the new architecture body.

```
architecture simple of dac is
-- function bit_to_real unchanged
  quantity vout across
           iout through aout; --
           branch quantities to ground
  signal dreal: real;
begin
  dreal <= bit_to_real(din);
  vout == vmax * dreal'ramp;
  -- infinite slope
end architecture simple;
```

Example 4. Generic D/A converter model revisited with S'Ramp (architecture only).

4.3. Structural Hierarchy: Analog Phase Locked Loop

The analog PLL model given in Example 5 is a VHDL 1076.1 structural model that instantiates three components: a phase detector, a lowpass filter, and a voltage-controlled oscillator. Each component has in turn its own model that describes its specific behavior. To be complete, this model would require a configuration declaration (not shown here) to bind the component instances in the netlist to their actual design entities.

```
library Disciplines;
use Disciplines.electrical_system.all;
entity apll is
  generic (fref: real := 1.0e6;
           -- reference frequency
           vref: real := 0.0;
           -- output voltage when
           f = fref
           slope: real := 5.0);
           -- in KHz/volt
  port (terminal pllin, pllout,
                 ref: electrical);
end entity apll;
architecture netlist of apll is
  component phase_detector is
    port (terminal in1, in2, outp,
                   ref: electrical);
  end component phase_detector;
  component lowpass is
    generic (gain, fp: real);
    port (terminal inp, outp,
                   ref: electrical);
  end component lowpass;
  component vco is
    generic (fref, slope, vref: real);
    port (terminal cp, cm,
                   p, m: electrical);
  end component vco;
  terminal t1, t2: electrical;
  -- local terminals
begin
  C_PD: phase_detector
    port map (in1 => pllin, in2 =>
              pllout, outp => t1,
              ref => ref);
  C_LP: lowpass
    generic map (gain => 20.0,
                 fp => fref/10.0)
    port map (inp => t1, outp => t2,
              ref => ref);
  C_VCO: vco
```

```vhdl
  generic map (fref => fref, slope =>
               slope, vref => vref)
  port map (cp => t2, cm => ref, p =>
            pllout, m => ref);
end architecture netlist;
```

Example 5. Analog PLL structural model.

The entity declaration is again straightforward. The architecture body declares the three components to instantiate and two local terminals *t1* and *t2* for local interconnections. Then, three component instantiation statements actually place the components in the netlist, define the actual values of generic parameters with generic maps, and define the interconnections with port maps. The explicit named associations allow for a better documentation.

The phase detector is implemented as a mixer that multiplies the sinusoidal PLL input by the output of the VCO (Example 6).

```vhdl
library Disciplines;
use Disciplines.electrical_system.all;
entity phase_detector is
  port (terminal in1, in2, outp,
        ref: electrical);
end entity phase_detector;
architecture mixer of phase_detector is
  quantity vout across iout through
           outp to ref;
  quantity vin1 across in1;
  quantity vin2 across in2;
begin
  vout == vin1 * vin2;
end architecture mixer;
```

Example 6. Phase detector model.

The lowpass filter is a single-pole filter that is implemented as a Laplace transfer function (Example 7). The coefficients of the numerator and the denominator are defined as constant vectors *num* and *den* whose element values are computed from the generic parameters *gain* and *fp*. The type *real_vector* is a new predefined type in VHDL 1076.1. The constitutive equation of the filter is that of an ideal voltage source whose value is the filtered input.

```vhdl
library IEEE, Disciplines;
use Disciplines.electrical_system.all;
use IEEE.math_real.all;
-- allows access to constant MATH_PI
entity lowpass is
```

```vhdl
  generic (gain, fp: real);
  -- filter gain and filter pole frequency
  port (terminal inp, outp,
        ref: electrical);
end entity lowpass;
architecture laplace of lowpass is
  quantity vin across inp;
  quantity vout across iout through outp;
  constant wp: real := 2.0*MATH_PI*fp;
  constant num: real_vector
                := (gain*wp);
  constant den: real_vector
                := (wp, 1.0);
begin
  vout == vin'Ltf(num, den);
  -- ideal voltage source
end architecture laplace;
```

Example 7. Lowpass filter model.

Finally, the voltage-controlled oscillator is described in Example 8. The break statement keeps the phase within the range $0.0\ldots 2\pi$. A discontinuity occurs when the phase becomes greater than 2π (i.e. when the quantity phase crosses the threshold value MATH_2_PI). The analog solver is then told to recompute the state of the model with the new initial condition *phase* **mod** MATH_2_PI for quantity *phase*.

```vhdl
library IEEE, Disciplines;
use Disciplines.electrical_system.all;
use IEEE.math_real.all;
entity vco is
  generic (fref, slope, vref: real);
  port (terminal cp, cm,
                 p, m: electrical);
end entity vco;
architecture bhv of vco is
  quantity verr across cp to cm;
  quantity vout across iout through
           p to m;
  quantity phase: real;
begin
  break phase => phase mod MATH_2_PI
    when phase'above(MATH_2_PI);
  phase'dot == MATH_2_PI * fref +
               slope * (verr - vref);
  vout == sin(phase);
end architecture bhv;
```

Example 8. VCO model.

4.4. *Noise Modeling: Noisy Diode*

The model of Example 1 is now extended to allow the noise simulation of the device model. Example 9 defines both shot noise and flicker noise through the declaration of two noise source quantities *q_shot* and *q_flicker*. Two new generic parameters *KF* and *AF* are also introduced, as the flicker noise coefficient and the flicker noise exponent, respectively. The contributions of the two noise source quantities are then added to the ideal diode current *id*. Both sources are zero (open-circuited) in a time domain simulation. They are only active when a noise simulation is running. Note that the flicker noise source quantity makes reference to the predefined function FREQUENCY that returns the current simulation frequency.

```
library IEEE, Disciplines;
use Disciplines.electrical_system.all;
use IEEE.math_real.all;
entity diode is
 generic (ISS, N, VT, TT, CJ0, VJ, RS,
          KF, AF: real);
 port (terminal p, m: electrical);
end entity diode;
architecture noisy of diode is
 constant elec_charge: real
          := 1.602e-19; -- in C
 quantity vd across id, ic through
          p to m;
 quantity q_shot: real
                  noise sqrt
                  (2.0*elec_charge*id);
 quantity q_flicker: real
                     noise sqrt(KF*id**
                     AF/frequency);
begin
 id == ISS * (exp((vd-RS*id)/(N*VT)) -
       1.0) + q_shot + q_flicker;
 ic'integ == TT*id -
             2.0*CJ0 * sqrt(VJ**2 -
             VJ*vd);
end architecture noisy;
```

Example 9. Diode model with shot noise and flicker noise.

5. Concluding Remarks

In this paper, the main aspects related to the future VHDL 1076.1 standard, an extension of VHDL 1076 to support the description and the simulation of analog and mixed-signal systems, have been presented. Thanks to solid mathematical foundations and a powerful set of constructs, the 1076.1 language applies for the modeling of more general systems than only electrical or electronic systems. Its capabilities for describing structure and behavior in a very general way allow it to support the modeling of both conservative and non-conservative systems at several levels of abstraction: at the functional (or signal-flow) level, where the system is described as a set of isolated functions that represents non-conservative input-output relationships (transfer functions), at the behavioral level where the system is described as arbitrary complex functions with conservative semantics, and at the circuit level where the system is described with basic constitutive functions that have a direct correspondence with physical devices, e.g. resistors, capacitors, sources, semiconductor devices. In addition, the language also naturally supports macromodeling techniques as it is done for example at the circuit level [14]. VHDL 1076.1 supports mixed-signal time domain simulation, and frequency domain simulation (small-signal AC and noise) for pure analog descriptions. It is not restricted in the kind of simulation algorithms the analog solver may use. Actually, great care has been taken to make language definitions independent of any particular implementation.

The development of the analog and mixed-signal extensions to VHDL has a long road behind it and includes contributions from many people. Language design is a complex task that requires to assemble many pieces in a coherent and consistent way. In the design of the VHDL 1076.1 language, the expertise came from analog design, mixed-signal design and VHDL language definition. At the time of this writing, a draft 1076.1 language reference manual is undergoing an IEEE balloting process to make the language the IEEE Standard 1076.1.

Acknowledgments

The material presented in this paper is the result of a collective effort of members from the IEEE 1076.1 Language Design Committee. By no means it is the result of the work of a single author, as this paper may incorrectly suggest.

DATE DUE

| ILL 4585585 | 12-10-99 |